Springer Laboratory

Springer Laboratory Manuals in Polymer Science

Schärtl, W.: Light Scattering from Polymer Solutions and Nanoparticle Dispersions
ISBN: 3-540-71950-4

Stribeck, N.: X-Ray Scattering of Soft Matter
ISBN: 3-540-46488-4

Koetz, J., Kosmella, S.: Polyelectrolytes and Nanoparticles
ISBN: 3-540-46381-X

Heinze, T., Liebert, T., Koschella, A.: Esterification of Polysaccharides
ISBN: 3-540-32103-9

Mächtle,W., Börger, L.: Analytical Ultracentrifugation of Polymers and Nanoparticles
ISBN: 3-540-23432-2

Brummer, R.: Rheology Essentials of Cosmetics and Food Emulsions
ISBN: 3-540-25553-2

Hatada, Kitayama: NMR Spectroscopy of Polymers
ISBN: 3-540-40220-9

Kulicke, Clasen: Viscosimetry of Polymers and Polyelectrolytes
ISBN: 3-540-40760-X

Pasch, Schrepp: MALDI-TOF Mass Spectrometry of Synthetic Polymers
ISBN: 3-540-44259-6

Mori, Barth: Size Exclusion Chromatography
ISBN: 3-540-65635-9

Pasch, Trathnigg: HPLC of Polymers
ISBN: 3-540-61689-6 (hardcover)
ISBN: 3-540-65551-4 (softcover)

Wolfgang Schärtl

Light Scattering from Polymer Solutions and Nanoparticle Dispersions

With 95 Figures and 16 Tables

Springer

Wolfgang Schärtl
Universität Mainz
Institut für Physikalische Chemie
Jakob-Welder-Weg 11
55099 Mainz
Germany
e-mail: schaertl@uni-mainz.de

Library of Congress Control Number: 2007924911

DOI 10.1007/978-3-540-71951-9

ISBN 978-3-540-71950-2 Springer Berlin Heidelberg NewYork

ISBN 978-3-540-71951-9 (eBook)

Springer is a part of Springer Science+Business Media

springer.com

Typesetting & Production: LE-TEX, Jelonek, Schmidt & Vöckler GbR, Leipzig, Germany
Coverdesign: WMXDesign GmbH, Heidelberg, Germany

SPIN 11530367 2/3180/YL – 5 4 3 2 1 0 Printed on acid-free paper

Springer Laboratory Manuals in Polymer Science

To my parents Marga and Anton Schärtl
without whom, in many respects,
this book would not have been written.

Preface

Light scattering is a very powerful method to characterize the structure of polymers and nanoparticles in solution. Recent technical developments have strongly enhanced the possible applications of this technique, overcoming previous limitations like sample turbidity or insufficient experimental time scales. However, despite their importance, these new developments have not yet been presented in a comprehensive form. In addition, and maybe even more important to the broad audience, there is the lack of a simple-to-read textbook for students and nonexperts interested in the basic principles and fundamental applications of light scattering. As part of the Springer laboratory handbook series, this book tries not only to provide such a simple-to-read and illustrative textbook about the seemingly very complicated topic "light scattering from polymers and nanoparticles in dilute solution," but also intends to cover some of the newest stateofthe-art technical developments in experimental light scattering.

It is a pleasure to acknowledge my sister Dr. Sabine Schärtl and several of my colleagues at Mainz University who have read parts of this book and offered criticism and helpful comments. I am especially indebted to Dr. Franziska Gröhn and Dr. Karl Fischer as well as to Waltraut Mueller and Christian Scherer for their careful study of the manuscript and their valuable suggestions. Last but not least, I would like to thank Professor Manfred Schmidt for encouraging me to write this book, and for the deeper insight into the light scattering method he helped me gain during the last 10 years.

Mainz, Germany
November 2006

Wolfgang Schärtl

Table of Contents

Glossary of Important Symbols and Abbreviations

A_2	second Virial coefficient
b^I, K	single particle scattering power or contrast factor
CUM	Cumulant data analysis of dynamic light scattering data
D_{app}	apparent selfdiffusion coefficient
d_f	fractal dimension
DLS	dynamic light scattering
dn_D/dc	refractive index increment
D_R	rotational selfdiffusion coefficient
D_s	selfdiffusion coefficient (translation)
$\langle D_s \rangle_z$	z-average selfdiffusion coefficient
D_T	translational selfdiffusion coefficient
E	electric field strength
f_c	coherence factor
FFF	field flow fractionation
FOQELS	fiber optic quasielastic light scattering
$F_s(q,\tau)$, $g_1(q,\tau)$	amplitude correlation function or dynamic structure factor
$g_2(q,\tau)$	intensity autocorrelation function
GPC	gel permeation chromatography
$G_s(r,\tau)$	van Hove selfcorrelation function
I_s	scattered intensity
L_k	Kuhn length
m	electric dipole momentum
MALLS or MALS	multiangle laser light scattering
MSCS	multispeckle correlation spectroscopy
M_w	weight average molar mass
n_D	index of refraction
$P(q)$	particle form factor
q	magnitude of scattering vector
R	absolute scattered intensity or Rayleigh ratio
R_g, s	radius of gyration
$\langle R_g^{\,2} \rangle_z$	z-average squared radius of gyration
R_H	hydrodynamic radius
$\langle R_H^{\,-1} \rangle_z$	inverse z-average hydrodynamic radius
SLS	static light scattering
α	polarizability

Γ	decay rate (of correlation function)
Γ_0, κ_1	first Cumulant ($=\langle D_s \rangle \cdot q^2$)
Γ_1, κ_2	second Cumulant
$\langle \Delta R(\tau)^2 \rangle$	mean-square particle displacement
η	solvent viscosity
θ	scattering angle
λ	wavelength of incident (and scattered) light
ρ	ρ -ratio ($= R_g / R_H$)
σ_R	particle size polydispersity
τ	correlation time

1 Fundamental Concepts

1.1 Introduction

All matter consists of atoms, which themselves are built from negative and positive charges. To describe the interaction of light with matter, one has to consider that light has both particular and wave character. Treating light within the classical wave picture, it is fairly simple to understand the origin of the phenomenon of light scattering: as an electromagnetic wave (a periodic modulation of electric and magnetic field strength both in space and time (see Eq. 1.1) for the electric field strength of a linearly polarized light beam of wavelength λ, propagating in x-direction), light will interact with the charges constituting a given molecule in remodelling the spatial charge distribution.

$$E(x,t) = E_0 \left(\sin\left(\frac{2\pi x}{\lambda}\right) + \sin\left(\frac{2\pi \frac{c}{\lambda}}{t}\right) \right) \tag{1.1}$$

The magnitude of this effect is given by a certain physical quantity: the polarizability of the molecule, that is, the ease of shifting charges within the molecule. The charge distribution follows the time-modulation of the electric wave vector of the incident light beam, and therefore the molecule constitutes an oscillating dipole or electric oscillator. This oscillating dipole acts as an emitter of an electromagnetic wave of the same wavelength as the incident one (for this reason, the process is called “elastic scattering”), emitted isotropically in all directions perpendicular to the oscillator as illustrated in Fig. 1.1.

The angle of observation with respect to the direction of the incident light beam is called the scattering angle and provides, as we will see further below, a measure for the length scale observed in a light scattering experiment.

For molecules or particles larger than 20 nm, several of these oscillating dipoles are created simultaneously within one given particle. As a consequence, some of the emitted light waves possess a significant phase difference. Accordingly, interference of the scattered light emitted from such an individual particle of size larger than 20 nm leads to a nonisotropic angular dependence of the scattered light intensity. The interference pattern of intraparticular scattered light, also called particle form factor, is characteristic for size and shape of the

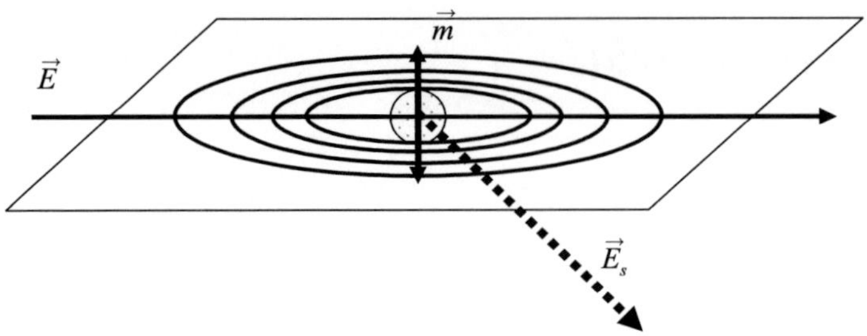

Fig. 1.1. Oscillating dipole induced by an incident light wave, and accordingly emitted light.

scattering particle. As a consequence, it provides the quantitative means for the characterization of particles in very dilute solution by light scattering. For particles smaller than $\lambda/20$, only a negligible phase difference exists between light emitted from the various scattering centers within the given particle. In this case, the detected scattered intensity will be independent of the scattering angle and only depend on the mass of the particle which is proportional to the total number of scattering centers one particle contains. The difference in the interference pattern of light scattered by very small and by larger particles, leading to a characteristic angular dependence of the measured scattered intensity for the latter, is illustrated in Fig. 1.2.

So far, we have considered light scattering as a purely elastic process where the emitted light has exactly the same wavelength as the incident light. Particles in solution, however, usually show a random motion (Brownian motion) caused by thermal density fluctuations of the solvent. As a consequence of the temporal changes in interparticle positions and the corresponding temporal concentration fluctuations, the interference pattern and the resulting scattered intensity detected at a given scattering angle also change with time, reflecting the Brownian motion of the scattering particles, as illustrated in Fig. 1.3.

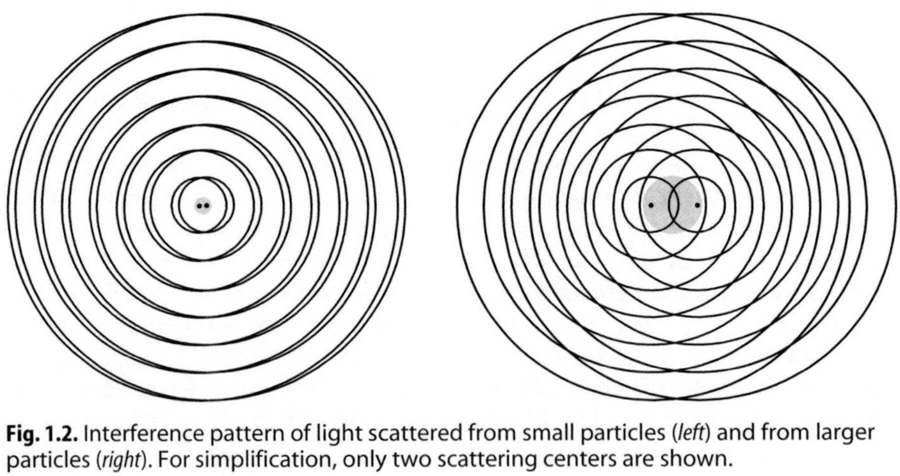

Fig. 1.2. Interference pattern of light scattered from small particles (*left*) and from larger particles (*right*). For simplification, only two scattering centers are shown.

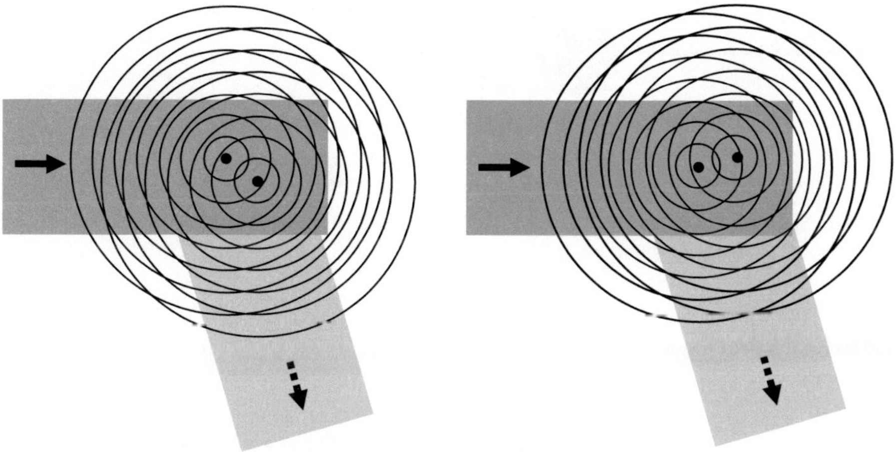

Fig. 1.3. Sketch of the change in the interference pattern of scattered intensity with time, caused by Brownian motion of two scattering particles.

This phenomenon provides the basis for dynamic light scattering, an experimental procedure which yields a quantitative measure for the mobility of scattering particles in solution as characterized by their selfdiffusion coefficient. Most modern particle sizers, frequently used both in industry and academia nowadays to determine the (hydrodynamic) size of particles in solution, are based on this principle.

In the following two chapters, the theoretical background of the two fundamental light scattering methods, that is static and dynamic light scattering, will be presented in more detail. A few mathematical relations which are most essential for the practice of light scattering will be highlighted in grey boxes to stress their importance.

1.2 Static Light Scattering

As mentioned above, matter scatters electromagnetic waves (light) due to the induction of an oscillating electric dipole, which serves as a source for the scattered light wave. The electric dipole momentum $\vec{m}$ depends on polarizability α and electric field vector $\vec{E}$ of the incident radiation as:

$$\vec{m} = \alpha \vec{E} \quad , \quad \left|\vec{E}\right| = E_0 \exp\left(i\left(2\pi\nu t - kx\right)\right) \tag{1.2}$$

$\nu = c/\lambda$ is the frequency of light of wavelength λ, and $\left|\vec{k}\right| = k = 2\pi/\lambda$ the length of the wave vector. In Eq. 1.2 I have assumed linearly and, in respect to the scattering plane (see below), vertically polarized light propagating into x-direction. The

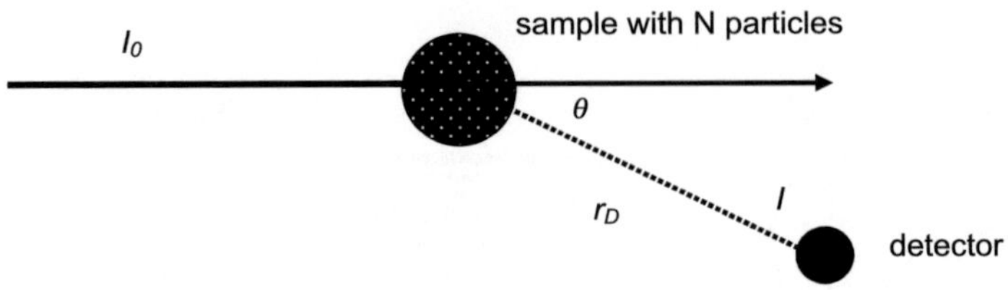

Fig. 1.4. Sketch of the light scattering process, including detection of the scattered intensity at scattering angle θ .

electric field vector of the scattered light wave emitted by the oscillating dipole is given as:

$$E_s = \left(\frac{\partial^2 m}{\partial t^2}\right)\frac{1}{r_D c^2} = \frac{-4\pi^2 v^2 \alpha E_0}{r_D c^2}\exp\left(i\left(2\pi v t - \vec{k}\vec{r}_D\right)\right) \tag{1.3}$$

Here, $\vec{r}_D$ is the distance vector from the scattering sample to the detector. In a light scattering experiment not the electric wave vector amplitude, but the scattered intensity $I_s = E_s E_s^* = |E_s|^2$ is detected. The principle setup for any static light scattering experiment, defining the important quantities, is illustrated in Fig. 1.4.

I_0 is the intensity of the incident light beam, I_s the intensity of the scattered light, θ the scattering angle and r_D the distance between sample and detector. The polarizability α depends on the dielectric permittivity ε (and correspondingly on the index of refraction n_D) as:

$$4\pi \frac{N}{V}\alpha = \varepsilon - 1 = n_D^{\ 2} - 1 \tag{1.4}$$

with N the number of scattering particles within the scattering volume V. The scattering volume is defined by the intersection of incident light beam and optical aperture used for observing the scattered light intensity, and therefore depends on the scattering angle as shown in Fig. 1.5.

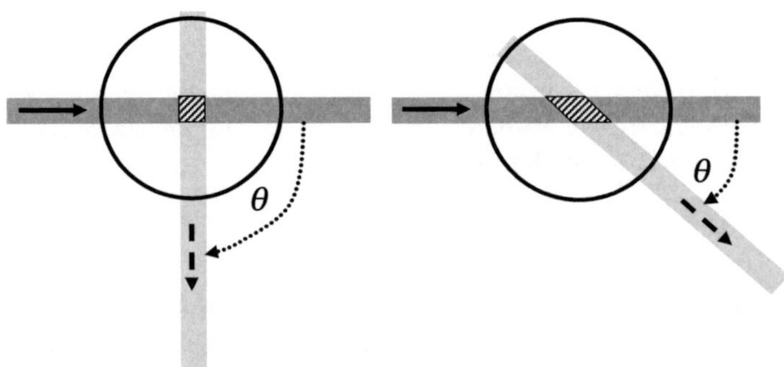

Fig. 1.5. Dependence of the scattering volume on the scattering angle.

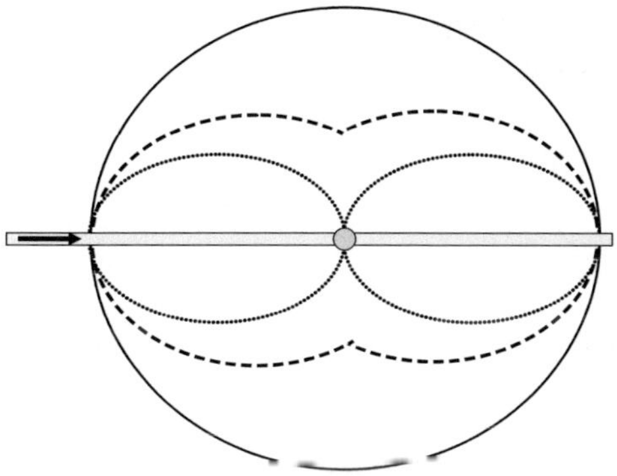

Fig. 1.6. Angular dependence for the polarization factors of light scattered from a solution of very small molecules, as detected in the scattering plane for different polarizations of the incident light wave (horizontal polarization ($2\ \cos^2\theta$, *dotted line*), nonpolarized light ($1+\cos^2\theta$, *dashed line*) and vertical polarization (2, *solid line*). The *grey bar* is the incident light beam, the *grey circle* in the center the scattering volume.

As a consequence, in a light scattering experiment the detected scattered intensity has to be normalized to a constant, scattering-angle-independent, scattering volume by factorization with $\sin\theta$.

Importantly, the polarization direction of the incident laser beam with respect to the scattering plane may cause an additional dependence of the detected scattered intensity on scattering angle: horizontally polarized light (h), for example, leads to a factor of $2\cdot\cos^2\theta$ for the scattered intensity. In case of unpolarized light (u), the polarization factor assumes a value of $1+\cos^2\theta$. The origin of this effect is that the intensity of the electromagnetic wave emitted by an oscillating dipole is strongest perpendicular to the axis of its oscillations, which in case of horizontally polarized light corresponds to a scattering angle $\theta=0°$. For simplification and if not stated otherwise, I will restrict myself in the following to vertical polarization (v) of the incident laser beam. In this case the polarization factor assumes a constant value of 2, and therefore the scattered intensity per constant scattering volume is independent of the scattering angle for very small particles of size <10 nm. The three different polarization factors for v, h, and u polarization of the incident laser light are illustrated in Fig. 1.6.

This sketch is valid for the absence of depolarization effects, as for example caused by rotation of optically anisotropic scattering molecules (see Chap. 5), in which case the polarization of the scattered light is identical to that of the incident light. Obviously, the maximum scattered intensity over the whole range of scattering angles is given for vertical polarization of the incident light, which is the reason why this scattering geometry is preferred in the experimental practice of light scattering.

1.2.1 Scattering from Dilute Solutions of Very Small Particles

For very dilute solutions of small scattering particles (for example nanoparticles or polymer chains of size smaller than $\lambda/20$, see above), the scattering intensity is independent of the scattering angle and, in case scattering from the density fluctuations of the solvent itself is ignored, only depends on the scattering power of the dissolved particles b, their mass concentration c and the osmotic pressure π:

$$I \sim b^2 kT \frac{c}{(\frac{\partial \pi}{\partial c})_{T,N}} \tag{1.5}$$

The interested reader should note that Eq. 1.5 has been derived from fluctuation theory, using $\left(\frac{d\mu}{dc}\right) = \frac{M_0}{\rho_0}\left(\frac{d\pi}{dc}\right)$ (with μ the chemical potential of the solvent in the solution, M_0 the molar mass of the solvent molecules and ρ_0 the solvent density).

According to van´t Hoff:

$$\frac{\partial \pi}{\partial c} = \frac{kT}{M} \text{ (ideal solutions)} \tag{1.6a}$$

$$\frac{\partial \pi}{\partial c} = kT(\frac{1}{M} + 2A_2 c + ...) \text{ (real solutions)} \tag{1.6b}$$

with M the molar mass of the dissolved particles and A_2 the second Virial coefficient which provides a quantitative measure for the solvent-solute-interactions. Here, "ideal solution" means the absence of specific interactions between solvent and solute molecules ($A_2 = 0$).

The scattering power b^2 (see Eq. 1.5) depends on the difference in polarizability of solute and solvent ($\Delta\alpha$), which itself depends on the respective refractive indices as:

$$\Delta\alpha = \alpha - \alpha_0 = \frac{\varepsilon - \varepsilon_0}{4\pi N/V} = \frac{n_D{}^2 - n_{D,0}{}^2}{4\pi N/V} = \frac{n_D{}^2 - n_{D,0}{}^2}{4\pi n} \tag{1.7}$$

n_D is the refractive index of the solute, $n_{D,0}$ the refractive index of the solvent and n the particle number density.

Using the refractive index increment

$$(\frac{\partial n_D}{\partial c}) \simeq \frac{n_D - n_{D,0}}{c} \tag{1.8}$$

the scattering power of one individual solute particle (b^2), also called contrast factor K, can be expressed as:

$$b^2 = \frac{4\pi^2}{\lambda_0{}^4 N_L} n_{D,0}{}^2 (\frac{\partial n_D}{\partial c})^2 = K \text{ in cm}^2\text{g}^{-2}\text{Mol} \tag{1.9}$$

Importantly, the scattered intensity scales inversely with the wavelength of the incident light to the power of four. Rayleigh scattering from the small gas molecules of our atmosphere is the reason that the sky looks blue: the short wavelength blue part of the spectrum of the incident sun light is scattered more strongly than the longer wavelength red part.

It should be noted that the scattered intensity given in Eq. 1.5 still depends on the actual experimental setup (for example on the sample-detector distance). This leads us to the so-called Rayleigh ratio *R*, which is an absolute scattering intensity not depending on the experimental conditions such as scattering volume *V* or the sample-detector distance r_D:

$$R = b^2 \cdot \frac{cM}{N_L} = \frac{4\pi^2}{\lambda_0^4} n_{D,0}^2 \left(\frac{\partial n_D}{\partial c}\right)^2 \frac{cM}{N_L} = (I_{solution} - I_{solvent}) \frac{r_D^2}{V} \qquad (1.10)$$

In practice, the absolute scattered intensity *R* of the solute particles is determined from the experimentally measured scattered intensities of the solution $I_{solution}$ and of the solvent $I_{solvent}$ as well as the intensity measured for a scattering standard I_{std} (typically the pure solvent toluene), renormalized by the so-called absolute scattering intensity of the standard $I_{std,abs}$:

$$R = (I_{solution} - I_{solvent}) \cdot \frac{I_{std,abs}}{I_{std}} \qquad (1.11)$$

Note that in this case all scattered intensities have to be measured using the same experimental setup. $I_{std,abs}$, the absolute scattering intensity of the standard (measured in [cm^{-1}]), can be found in reference tables (see also Chap 6 of this book).

Importantly, the reader should keep in mind that for comparatively small particles of size <10 nm, in which case intraparticular interference of the scattered light becomes negligible, the absolute scattered intensity of a given sample detected at any scattering angle is constant and only depends on the optical contrast b^2 and the number of particles within the scattering volume (= number density $n = N/V$) times the mass of a single scattering particle squared, $R = b^2 \cdot cM/N_L = b^2 \cdot n \cdot M^2/N_L^2$. This is simple to understand since the number of individual scattering centers of a single scattering particle is given by the molar mass of the particle. Therefore, the scattered intensity has to be proportional to the particle mass squared. This point has been stressed here because it is needed to determine which type of average sample characteristics is obtained by light scattering from polydisperse samples containing small particles, like for example synthetic polymer chains: the most important of these average sample characteristics determined in a static light scattering measurement are the z-average radius of gyration and the weight-average molar mass, as will be explained in detail further below.

Finally, for nonideal solutions (see Eq. 1.6b above), Eq. 1.10 can be rewritten to yield the basic equation for static light scattering experiments on solutions of small (size <10 nm) particles:

$$\frac{Kc}{R} = \frac{1}{M} + 2A_2 c + \ldots \qquad (1.12)$$

1.2.2 Scattering from Dilute Solutions of Larger Particles

For larger scattering particles, the scattered intensity is no longer independent of the scattering angle. The so-called scattering vector $\vec{q}$ (in [cm^{-1}]), which is experimentally determined by the scattering angle θ and the wavelength of the laser light λ. provides a quantitative measure for the length scale of the static light scattering experiment. Figure 1.7 shows how the value of $\vec{q}$ is derived from a given scattering geometry:

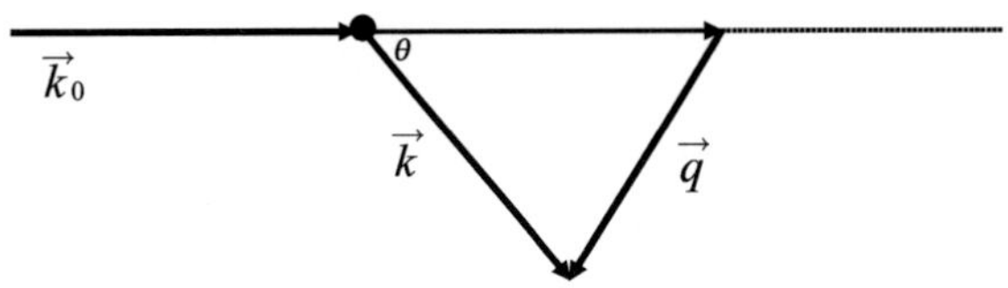

Fig. 1.7. Sketch of the definition of the scattering vector $\vec{q}=\vec{k}-\vec{k}_0$.

$\vec{k}_0$ and $\vec{k}$ are the wave vectors of the incident and of the scattered light beam; θ is the scattering angle. The scattering vector $\vec{q}$ is simply the difference of the two wave vectors, i.e., $\vec{q}=\vec{k}-\vec{k}_0$. For an elastic scattering process, $\left|\vec{k}_0\right|=\left|\vec{k}\right|=2\pi/\lambda$, and therefore:

$$\left|\vec{q}\right|=q=\frac{4\pi\sin(\theta/2)}{\lambda} \tag{1.13}$$

In addition, the refractive index of the solvent itself n_D has to be taken into account since it changes the wavelength of the incident light compared to its value in air ($n_D=1$):

$$q=\frac{4\pi n_D\sin(\theta/2)}{\lambda} \tag{1.14}$$

For very dilute solutions, interferences between different scattering particles, the so-called structure factor, can be neglected. In this case, the angular dependence of the measured scattered intensity *I(q)* is only caused by intraparticular interferences, leading us by pair-wise summation over all scattering centers of a single particle to:

$$I(q)=N^2b^2\sum_{i=1}^{Z}\sum_{j=1}^{Z}\exp\left[-i\vec{q}\left(\vec{r}_i-\vec{r}_j\right)\right]=N^2b^2\sum_{i=1}^{Z}\sum_{j=1}^{Z}\exp\left[-i\vec{q}\vec{r}_{ij}\right] \tag{1.15}$$

Here, our sample consists of *N* identical particles within the scattering volume, each particle containing *Z* scattering centers. *i, j* stand for two of these scattering centers within the same particle, and $\vec{r}_{ij}$ are the distance vectors between them defining the particle density distribution, accordingly.

Radius of Gyration/Molar Mass/Zimm Equation

For isotropic particles the normalized single particle scattering, also called particle form factor $P(q)$, after series expansion is given as:

$$P(q)=\frac{1}{N^2Z^2b^2}I(q)=\frac{1}{Z^2}\sum_{i=1}^{Z}\sum_{j=1}^{Z}\exp\left(-i\vec{q}\vec{r}_{ij}\right)=\frac{1}{Z^2}\sum_{i=1}^{Z}\sum_{j=1}^{Z}\left(1-\frac{1}{6}q^2r_{ij}^{\,2}+\dots\right) \quad (1.16)$$

For simplification, I will introduce here the so-called center of mass coordinate system: the origin of the Cartesian coordinate system is transferred to the particle's center of mass, as shown in Fig. 1.8.

Note that Cartesian position vectors r_i accordingly have to be replaced by the center of mass-based position vectors ($\vec{s}_i$). If we assume a homogeneous particle with constant particle density $\rho\left(\vec{s}_i\right)=\rho$, we find:

$$\sum_{i=1}^{Z}\vec{s}_i=0 \text{ and } s^2=\frac{1}{Z}\sum_{i=1}^{Z}\vec{s}_i^{\,2}\neq 0 \quad (1.17)$$

s^2 is the squared radius of gyration of the scattering particle, usually denoted as $R_g^{\,2}$. Inserting the distance vector $\vec{r}_{ij}=\vec{s}_j-\vec{s}_i$, we can rewrite the summation expression contained in the particle form factor (Eq. 1.16):

$$\frac{1}{Z^2}\sum_{i=1}^{Z}\sum_{j=1}^{Z}\vec{r}_{ij}^{\,2}=\frac{1}{Z^2}\sum_{i=1}^{Z}\sum_{j=1}^{Z}(s_i^{\,2}+s_j^{\,2}-2(\vec{s}_i\vec{s}_j))=2s^2 \quad (1.18)$$

where we have used that $\sum_{i=1}^{Z}\sum_{j=1}^{Z}\vec{s}_i\vec{s}_j=0$ if $i\neq j$. Therefore, we finally get for the particle form factor:

$$P(q)=1-\frac{1}{3}s^2q^2+\dots \quad (1.19)$$

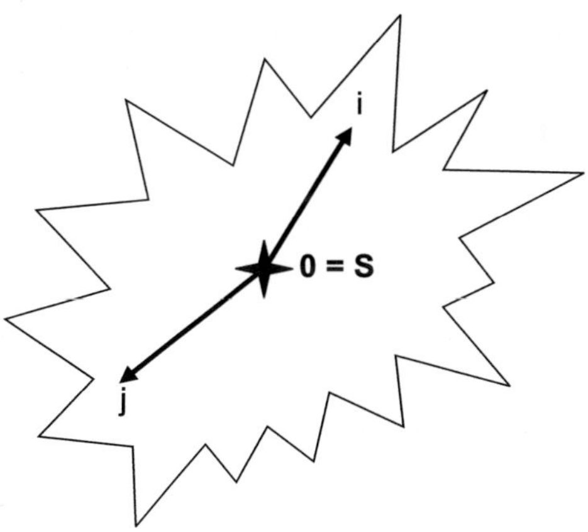

Fig. 1.8. Definition of the center-of-mass and the corresponding center-of-mass coordinates of a scattering particle.

To take into account the effects of particle concentration and solute-solvent interactions on the measured scattering intensity, we have to refer to thermodynamic fluctuation theory. In conclusion, it can be shown that the normalized absolute scattering intensity R depends on the particle form factor as:

$$Kc/R = 1/MP(q) + 2A_2c + \dots \tag{1.20}$$

Inserting $P(q)$ from Eq. 1.19 into Eq. 1.20, we derive the very important Zimm equation:

$$Kc/R = 1/M\left(1 + 1/3\, s^2q^2\right) + 2A_2c \tag{1.21}$$

This equation provides the basis for analyzing the scattered intensity from comparatively small particles ($s^2q^2 << 1$, in case of light scattering: 10 nm < particle radius <50 nm) to determine the molar mass, the radius of gyration $\langle s^2 \rangle^{1/2}$ and the second Virial coefficient A_2, the latter providing a quantitative measure for the solute particle-solvent interactions. Here, $\langle\ \rangle$ stands for the isotropic particle ensemble average, which is an orientational average in case of rod-like scattering particles, or a chain conformation average in case of random polymer coils. The reader should note that in a static light scattering experiment, even in case of very dilute sample solutions, an extremely large number of particles are located in the scattering volume (>10e9). Therefore, the measured particle form factor corresponds to an orientational average for anisotropic particles, which are randomly oriented within the sample. For linear polymer chains of identical chain length, on the other hand, each random polymer coil may assume a different conformation. In this case, the measured particle form factor corresponds to a conformational ensemble average. In addition, the orientation and/or conformation of a given scattering particle is changing with time.

So far, we have only considered so-called monodisperse particle solutions, that is, solutions which contain solute particles all identical in size and shape. For polydisperse samples, the Zimm analysis of the light scattering data according to Eq. 1.21 yields the following averages:

(i) The weight average of the molar mass

$$M_w = \frac{\sum_{k=1}^{K} N_k M_k M_k}{\sum_{k=1}^{K} N_k M_k} \tag{1.22}$$

(ii) The z-average of the squared radius of gyration:

$$< s^2 >_z = \left\langle R_g^{\,2} \right\rangle_z = \frac{\sum_{k=1}^{K} N_k M_k^{\,2} s_k^{\,2}}{\sum_{k=1}^{K} N_k M_k^{\,2}} \tag{1.23}$$

Here, our sample consists of K particle species of different molar mass and different size but identical chemical composition, and N_k is the number of scattering particles of species k [each with identical molecular mass M_k and identical squared radius of gyration (s_k^2)] within the scattering volume. It is straight forward to understand the origin of these averages keeping in mind what I described in some detail above: in the small particle regime considered here, the scattered intensity of a given particle species k depends on its optical contrast, which is identical for all particles irrespective of their molar mass, and on number and molar mass as $N_k M_k^2$. Therefore, sample characteristics measured in a light scattering experiment and depending directly on the scattered intensity correspond to so-called z averages (see Eq. 1.23), where the number distribution function (N_k) of the respective quantity has to be weighted with M_k^2. On the other hand, the reduced scattered intensity Kc/R (see Eq. 1.21) is determined from the ratio of sample mass concentration and measured scattered intensity, reducing the average of the molar mass to a weight average M_w, where the mass distribution $N_k M_k$ of the polydisperse sample is only weighted with M_k.

Particle Form Factor for "Large" Particles

Without series expansion (see Eq. 1.16), which is only valid for light scattering from small particles, the particle scattering form factor is given as:

$$P(q) = \frac{1}{Z^2} \sum_{i=1}^{Z} \sum_{j=1}^{Z} \exp(-i\vec{q}\vec{r}_{ij}) \tag{1.24}$$

For homogeneous spherical particles, one gets:

$$P(q) = \frac{9}{(qR)^6} \left(\sin(qR) - qR\cos(qR)\right)^2 \tag{1.25}$$

with R the radius of the sphere.This expression (Eq. 1.25) corresponds to an oscillating function, as shown in Fig. 1.9.

The position of the first minimum is found at $qR = 4.49$, which can be used to easily determine the particle radius R. Note that the oscillations are not as well pronounced for scattering from polydisperse spherical particles, as shown in Fig. 1.10.

For some other simple particle morphologies, the following particle form factors are obtained. Importantly, all form factors of anisotropic particles given here, like the form factors of thin disks or rigid rods, are orientational ensemble averages, as has been discussed above.

1. Hollow sphere of radius R with very thin shell:

$$P(q) = \left[\frac{\sin(qR)}{qR}\right]^2 \tag{1.26}$$

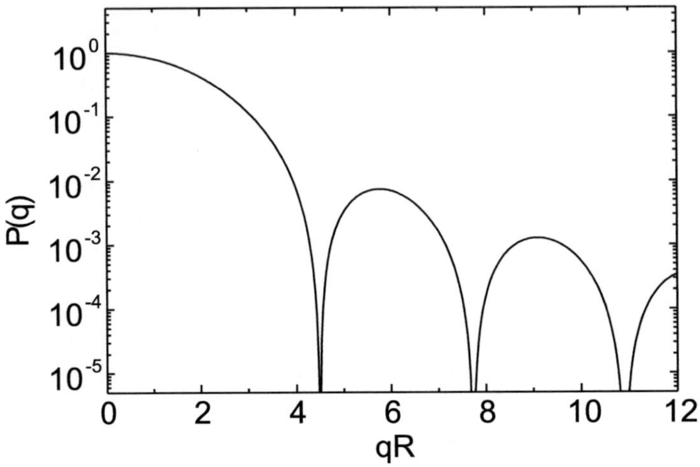

Fig. 1.9. Particle form factor for monodisperse spheres.

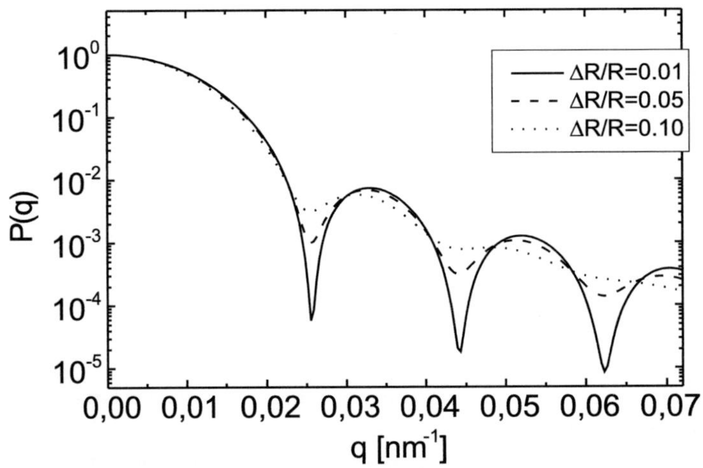

Fig. 1.10. Particle form factor for polydisperse spheres.

2. Thin disk of radius R:

$$P(q) = \frac{2}{(qR)^2}\left[1 - \frac{1}{qR} J_1(2qR)\right] \tag{1.27}$$

with J_1 being the so-called first-order Bessel function.

3. Thin cylinder of length L:

$$P(q)=\frac{2}{qL}Si(qL)-\frac{\sin^2\left(\frac{qL}{2}\right)}{\left(\frac{qL}{2}\right)^2} \tag{1.28}$$

with $Si(x)=\int_0^x \frac{\sin u}{u}du$ the so-called sinus-integral-function.

Next, I will briefly discuss an alternative approach towards determining the topology of solute particles from the angular dependence of the scattered intensity, besides fitting the experimental data to an assumed particle form factor according to Eqs. 1.26–1.28. This second method is based on the so-called fractal dimension, which is the scaling of particle mass with particle size.

Radius of Gyration and Geometrical Radius for Particles of Various Morphologies in Dependence of Molar Mass

In general, the radius of gyration is given as a volume integral over the mass distribution of a given particle, that is:

$$s^2=\frac{\int_V m(\vec{r}_i)r_i^2 dV}{\int_V m(\vec{r}_i)dV} \tag{1.29}$$

In the following, solutions of this equation for some selected particle morphologies will be presented:

1. For a homogeneous sphere of radius R and mass density ρ:

$$s^2=\frac{\int_0^R \rho 4\pi r^2 r^2 dr}{\int_0^R \rho 4\pi r^2 dr}=3/5\,R^2 \tag{1.30}$$

For the molar mass, one gets:

$$M^{2/3}\sim s^2\sim R^2 \tag{1.31}$$

2. For a hollow sphere of radius R and very thin shell d_s:

$$s^2=\frac{\rho\cdot 4\pi R^2 d_s R^2}{\rho\cdot 4\pi R^2 d_s}=R^2 \tag{1.32}$$

$$\text{and } M\sim s^2\sim R^2 \tag{1.33}$$

3. For a very thin disk of radius R and thickness d_s :

$$s^2 = \frac{\int_0^R \rho \cdot 2\pi r d_s r^2 dr}{\int_0^R \rho \cdot 2\pi r d_s dr} = \frac{1}{2} R^2 \tag{1.34}$$

$$\text{and } M \sim s^2 \sim R^2 \tag{1.35}$$

4. For a very thin cylinder of length L and thickness d:

$$s^2 = \frac{\int_{-L/2}^{L/2} \rho \cdot \pi \left(\frac{d}{2}\right)^2 r^2 dr}{\int_{-L/2}^{L/2} \rho \cdot \pi \left(\frac{d}{2}\right)^2 dr} = \frac{L^2}{12} \tag{1.36}$$

$$\text{and } M^2 \sim s^2 \sim L^2 \tag{1.37}$$

5. For an unperturbed polymer coil with Z segments each of Kuhn length l:

$$< s^2 > = \frac{1}{Z^2} \int_0^Z (Z - s) s l^2 ds = \frac{l^2 Z}{6} = \frac{< R^2 >}{6} \tag{1.38}$$

$$\text{and } M^{0.5} \sim \left\langle s^2 \right\rangle^{0.5} \sim Z^{0.5} \tag{1.39}$$

with $\left\langle R^2 \right\rangle$ the mean-square end-to-end-distance of the polymer coil. The $\langle\ \rangle$, as described before, corresponds to a conformational ensemble average, since not only the chain conformations of all polymer coils detected within the scattering volume are different, but also the conformation of a given particle is changing with time.

Fractal Dimensions

In general, for any particle of given topology its mass M scales with its size R according to:

$$M(R) \sim R^{d_f} \tag{1.40}$$

In Eq. 1.40, d_f is the so-called fractal dimension of the particle. Very thin cylinders, for example, have a fractal dimension $d_f = 1$, thin disks $d_f = 2$ and compact spheres $d_f = 3$ (see Eqs. 1.37, 1.35, and 1.31 above). In the experimental scattering vector regime $q > R_g^{-1}$, the scattered intensity of a fractal particle is given as:

$$I(q) \approx q^{-d_f}, \text{ or } \log I(q) = -d_f \cdot \log q \tag{1.41}$$

Table 1.1. Fractal dimensions of selected topologies

Topology	Fractal dimension d_f
Cylinders, rods	1
Ideal Gaussian coil	2
Gaussian coil with excluded volume	5/3
Branched Gaussian chain	16/7
Swollen branched chain	2
2D-objects with smooth surfaces	2
2D-objects with fractal surfaces	1–2
3D-objects with smooth surfaces	3
3D-objects with fractal surfaces	2–3

If $\log I(q)$ is plotted vs. $\log q$, one obtains a linear decay with slope d_f, allowing to directly determine the fractal dimension of the scattering particle. Here, it should be noted that this procedure works only in experimental practice if the measured scattered intensity shows this linear decay over at least one order of magnitude of the experimental q-regime. Therefore, the applicability of this scheme depends on the average size of the scattering particles in respect to the wavelength of the incident light as well as on the experimentally available scattering angles. For illustration, an experimental example from literature will be reviewed in Chap. 5 of this book in more detail. Table 1.1 summarizes the fractal dimensions of a selected number of various important particle topologies.

Concluding Remarks

(i) The magnitude of the scattering vector $\vec{q}$ defines the observational length scale of the light scattering experiment. The larger q becomes, the smaller the length scale and the sample details observed. In Fig. 1.11 and Table 1.2 the meaning of this fundamental concept of an experimental length scale defined by the scattering angle is illustrated for the exemplary case of a random polymer coil.

Table 1.2. q-scale vs. structural details of scattering particles observed, respectively (example: polymer coils in solution)

q-scale	Resolution	Information	Comments
$qR \ll 1$	Whole coil	Mass, radius of gyration	e.g., Zimm plot
$qR < 1$	Topology	Cylinder, sphere, …	
$qR \cong 1$	Topology quantitative	Size of cylinder, …	
$qR > 1$	Chain conformation	Helical, stretched, …	
$qR \gg 1$	Chain segments	Chain segment density	

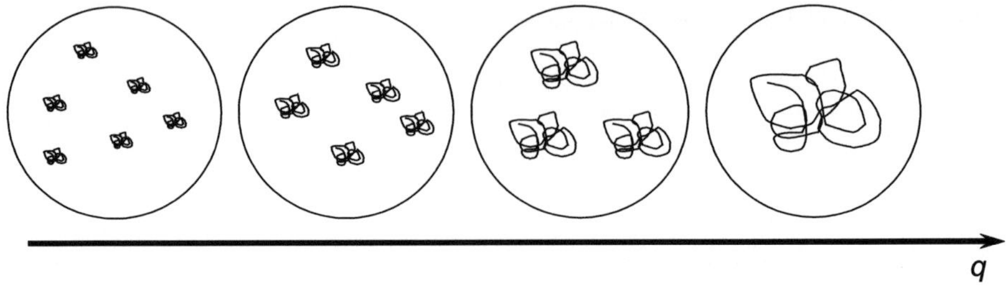

Fig. 1.11. q-scale vs. sample details observed in case of a random polymer coil (see also Table 1.2).

As shown, one may consider light scattering also as an inverse microscopic technique, and the scattering vector as the resolution or magnification of the inverse microscope: the smaller q, the lower the magnification and the corresponding resolution of the structure of a given scattering particle.

(ii) The formalism presented here is valid for light scattering studies of very dilute samples, in which case interactions between the scattering particles, leading to an interparticular order which causes the presence of a so-called static structure factor (= interparticle interferences of scattered light), can be neglected. Only for such dilute samples, the measured scattered intensity (normalized using an absolute scattering standard, see Eqs. 1.11, 1.20) represents the pure particle form factor defined in the fundamental equation given above (Eq. 1.24). The reader should note that, in case of charged scattering particles, sample concentrations even lower than 0.1 g/L may give rise to a structure factor due to the long-range Coulomb interparticle repulsion. Addition of salt to screen these unwanted Coulomb interactions can usually solve the problem. In Chaps. 4 and 5 of this book, I will review the problem of light scattering from charged systems in more detail.

1.3 Dynamic Light Scattering

As mentioned above, if the scattering particles are moving, fluctuations in the scattered intensity with time are directly reflecting the so-called Brownian particle motion of the scattering particles (caused by thermal density fluctuations of the solvent). This is the case because of a change in the interference pattern with changing interparticle position, and correspondingly a change in the detected scattered intensity measured at a given scattering angle (see Fig. 1.3).

To quantitatively analyze the particle mobility by light scattering, it is helpful to express the scattering intensity fluctuations in terms of correlation functions, as will be discussed in detail in this section.

Time-Intensity-Autocorrelation Function and Particle Motion

The dynamic structure factor $F_s(\vec{q},\tau)$ contains all information concerning the motion of the scattering solute particle. It is the Fourier transform of the so-called van Hove selfcorrelation function $G_s(\vec{r},t)$:

$$F_s(\vec{q},\tau) = \int G_s(\vec{r},\tau)\exp(i\vec{q}\vec{r})d\vec{r} \tag{1.42}$$

$$G_s(\vec{r},\tau) = < n(\vec{0},t)n(\vec{r},t+\tau) >_{V,T} \tag{1.43}$$

Here, $n(\vec{r},t)$ is the local number density of scattering particles (= number of scattering particles, fluctuating with time due to Brownian motion, within a very small sub volume of the scattering volume centered at position $\vec{r}$) at a given time *t*. In principle and for very dilute solutions, $G_s(\vec{r},\tau)$ defines the probability of finding a given scattering particle at time $t+\tau$ and position $\vec{r}$, if the same particle previously at time *t* has been located at position $\vec{0}$. It should be noted here that, for the dynamic scattering process, not the absolute position vectors $\vec{r}$ and $\vec{0}$ or the absolute times matter, but only the relative distance vector $\vec{r}-\vec{0}$ as well as the time difference τ are important. It plays no role where one arbitrarily chooses the origin of the coordinate system $\vec{0}$ or the starting time of the experiment. The average <> is taken both over the whole scattering volume and the total measuring time. For an isotropic diffusive particle motion (= Brownian motion), also called "random walk", $G_s(r,\tau)$ only depends on the distance $r=\left|\vec{r}\right|$, and is given as:

$$G_s(r,\tau) = [2\pi/3 < \Delta R(\tau)^2 >]^{3/2} \exp\left(-\frac{3r(\tau)^2}{2 < \Delta R(\tau)^2 >}\right) \tag{1.44}$$

with $\left\langle \Delta R(\tau)^2 \right\rangle$ the mean-square displacement of the scattering particle, that is, the average distance squared it travels during time τ. The Brownian particle motion is, as already mentioned, caused by random thermal density fluctuations of the solvent molecules which push the scattering particle along. The scattering particle therefore exhibits a random walk through the scattering volume, and the mean-square displacement is given as:

$$\left\langle \Delta R(\tau)^2 \right\rangle = 6D_s\tau \tag{1.45}$$

with D_s the selfdiffusion coefficient. Note that in this case the van Hove selfcorrelation function $G_s(r,\tau)$ is a Gaussian curve with its half width given by the diffusion coefficient (see Eqs. 1.44 and 1.45). Fourier transform leads to the corresponding dynamic structure factor, which is the primary quantity measured in the dynamic light scattering experiment:

$$F_s(q,\tau) = \exp\left(-q^2 < \Delta R^2(\tau) >_T \tau/6\right) = \exp\left(-D_s q^2 \tau\right) \tag{1.46}$$

The Stokes–Einstein–equation,

$$D_s = \frac{kT}{f} = \frac{kT}{6\pi\eta R_H} \tag{1.47}$$

allows one to determine the hydrodynamic radius R_H of the scattering particle, if sample temperature T and solvent viscosity η are known and the selfdiffusion coefficient is measured by dynamic light scattering. Note here that R_H is the radius of an equivalent sphere, experiencing during its Brownian motion in solution a friction f identical in magnitude to that of our scattering particle which itself is not necessarily a sphere. Let us now review the theoretical background of the dynamic light scattering experiment in more detail.

Theory of the Dynamic Light Scattering Experiment

$G_s(r,\tau)$, or $F_s(q,\tau)$ respectively, can either be determined experimentally by Fabry–Perot–interferometry (which is beyond the scope of this book) or by dynamic light scattering. The principle of how the fluctuating scattered intensity $I(q,t)$ is treated in a dynamic light scattering experiment is illustrated in Fig. 1.12.

At the top of Fig. 1.12, the signal detected by the photomultiplier at a given scattering angle is shown. For static light scattering experiments, the average scattered intensity $\langle I(q,t)\rangle_T$, as indicated by the dotted line, is measured. For dynamic light scattering, on the other hand, the detailed analysis of the fluctuating intensity $I(q,t)$ is important. For this purpose, the fluctuation pattern is "mathematically translated" into an intensity autocorrelation function, using a hardware correlator: the time-dependent scattered intensity is multiplied with itself after it has been shifted by a distance τ in time, and these products are averaged over the total measurement time. This intensity autocorrelation function $< I(q,t)I(q,t+\tau) >$, which is not depending on t but only on the correlation time τ, is calculated for various values of τ, ranging in a typical dynamic light scattering experiment from about 100 ns to several s. Here, the lower time limit is given by the detector hardware, and the upper correlation time is limited by the stability of the dynamic light scattering setup and the channel number of the hardware correlator (see examples reviewed in Chap. 5.1). For scattering particles in solution exhibiting simple Brownian motion, the intensity correlation function should decay exponentially from 2 to 1. It is related to the so-called dynamic structure factor (or amplitude autocorrelation function $F_s(q,\tau)$) via the Siegert relation:

$$F_s(q,\tau) = \exp(-D_s q^2 \tau) = < E_s(q,t)E_s{}^*(q,t+\tau) > = \sqrt{\frac{< I(q,t)I(q,t+\tau) >}{\langle I(q,t)^2\rangle} - 1} \tag{1.48}$$

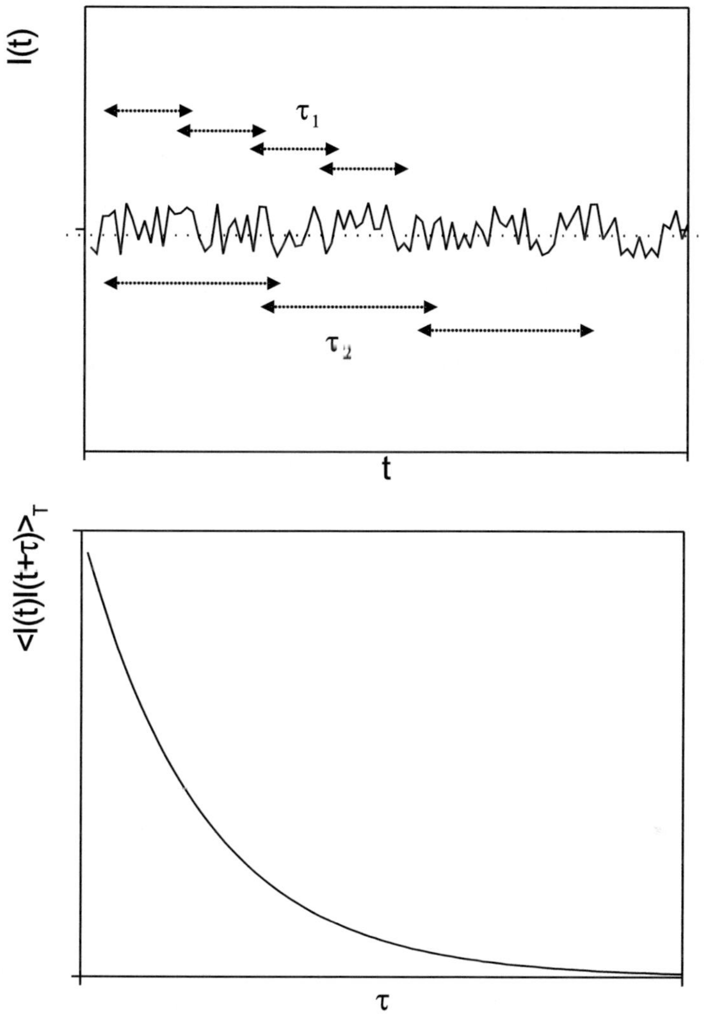

Fig. 1.12. Principle of dynamic light scattering: sketched are the intensity fluctuations and the procedure to calculate the intensity autocorrelation function shown at the bottom of the figure.

For monodisperse samples, $F_s(q,\tau)$ is a single exponential with decay rate $\Gamma = D_s q^2$ (or relaxation time $(D_s q^2)^{-1}$). Here, it should be noted that many authors use a different nomenclature compared to the one introduced so far. They call the normalized scattered intensity autocorrelation function, which theoretically should be a single-exponential decaying from 2 to 1, $g_2(q,\tau)$:

$$g_2(q,\tau) = \frac{< I(q,t)I(q,t+\tau) >}{\langle I(q,t)^2 \rangle} \tag{1.49}$$

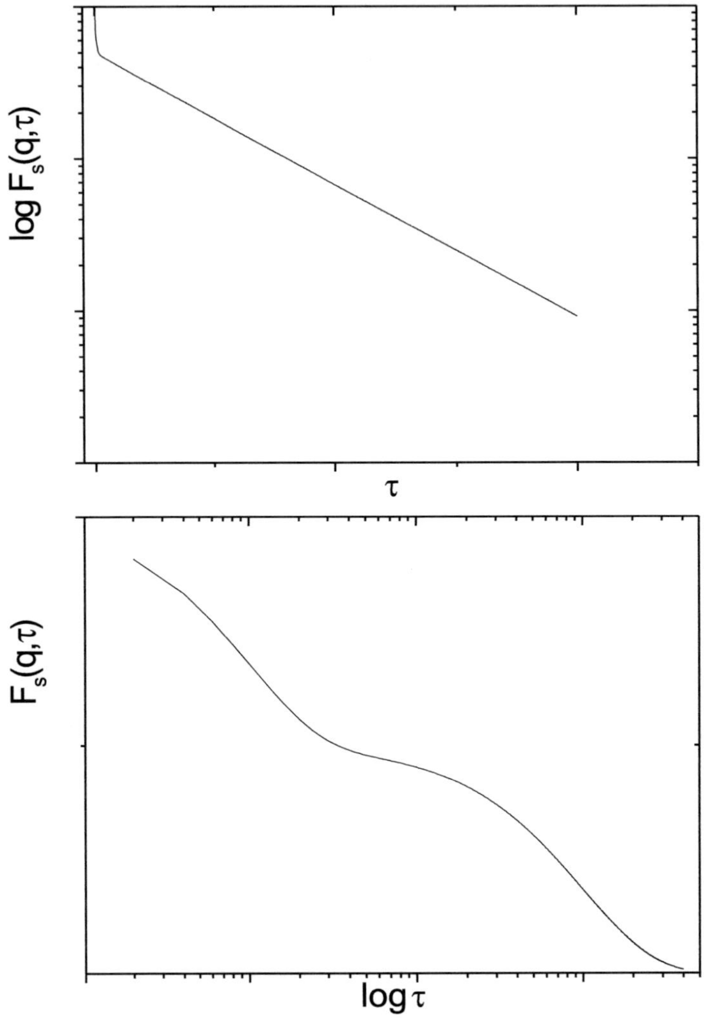

Figs.1.13. log-lin and lin-log plot of $F_s(q,\tau)$ for a bimodal sample.

In this nomenclature, the amplitude correlation function or dynamic structure factor is called $g_1(q,\tau)$, and correspondingly the Siegert relation is rewritten as:

$$g_2(q,\tau)=1+g_1(q,\tau)^2 \tag{1.50}$$

Typically, $F_s(q,\tau)$ (or $g_1(q,\tau)$) is plotted not in a linear scale but in a semi-logarithmic scale (log-lin and lin-log). This makes the data analysis much easier for polydisperse samples, as shown in Fig. 1.13 for a bimodal sample (a sample containing scattering particle species with two different sizes).

In a log-lin plot, the correlation function $F_s(q,\tau)$ is a single straight line for a monodisperse sample and a combination of two lines with different slopes for the bimodal example. In a lin-log plot, $F_s(q,\tau)$ shows a step-like decay where the number of steps reflects the number of different particle sizes (e. g., two steps in case of a bimodal sample, as shown in Fig. 1.13).

The selfdiffusion coefficient, and therefore the hydrodynamic radius of the scattering particles, can be determined by dynamic light scattering only in case of very dilute samples. In more concentrated samples, interactions between the scattering particles may have a strong influence on the particle mobility. Here, dynamic light scattering provides a powerful method to quantify interparticle interactions in solution. In this book, however, the focus will be mainly on particle characterization, as already has been stated above in the section on static light scattering. The reader should therefore keep in mind that all theoretical descriptions presented so far are only valid for very dilute scattering particle solutions, where interactions between the scattering solute particles have no influence on sample structure and/or particle mobility.

Finally, I should not forget to mention that, utilizing the polarization of incident and scattered light, so-called depolarization dynamic light scattering allows the measurement of the rotational Brownian motion (described quantitatively by a characteristic rotational diffusion coefficient) of anisotropic scattering particles, like nanorods or cylindrical micelles. A detailed discussion of the theory of rotational diffusion and the corresponding hydrodynamic friction terms is beyond the scope of this textbook, but some illustrative experimental light scattering examples, showing how rotational diffusion coefficients can be measured and how these data are interpreted, are presented in Chap. 5.1.

Dynamic Light Scattering from Polydisperse Samples

For polydisperse samples with size distribution $P(R_H)$, the experimentally determined (average) selfdiffusion coefficient is defined by a distribution function $P(D_s)$. Importantly, this distribution function depends not only on the particle number density of species i (n_i), but also on particle mass M_i and particle form factor $P_i(q)$, since the scattered intensity of a given particle species i is given as $I_i \sim n_i \cdot M_i^2 \cdot P_i(q)$. In this case, $F_s(q,\tau)$ is not a simple single exponentially decaying function, but a superposition of several such single exponentials weighted by $P(D_s)$ as:

$$F_s(q,\tau) = \int_0^\infty P(D_s)\exp(-q^2 D_s \tau)dD_s \tag{1.51}$$

Since the particle form factor contributes to $P(D_s)$, for scattering particles larger than 10 nm the measured diffusion coefficient distribution $P(D_s)$ does not only depend on the particle size distribution itself but also on the scattering vector q, as will be discussed in more detail further below. Note here that, in case of nonspherical scattering particles, not only polydispersity but also nontranslational particle motion like rotation or polymer segment fluctuations may cause a q-dependence of the measured apparent diffusion coefficient.

One way of analyzing the more complicated (compared to the case of monodisperse samples) data obtained from a polydisperse sample in a quantitative way is the so-called "Cumulant analysis." It is based on a series expansion of $F_s(q,\tau)$, and therefore is only valid for small size polydispersities $\Delta R_H / \langle R_H \rangle \leq 20\%$:

$$\ln F_s(q,\tau) = -\kappa_1 \tau + \frac{1}{2!}\kappa_2 \tau^2 - \frac{1}{3!}\kappa_3 \tau^3 + \ldots \qquad (1.52)$$

Here, the first Cumulant $\kappa_1 = \langle D_s \rangle q^2$ yields the average diffusion coefficient $\langle D_s \rangle$ and the corresponding inverse average hydrodynamic radius $\langle R_H^{-1} \rangle$. The second Cumulant $\kappa_2 = \left(\langle D_s^2 \rangle - \langle D_s \rangle^2\right) q^4$ provides a quantitative measure for the polydispersity of the diffusion coefficient distribution function (σ_D), which is defined as:

$$\sigma_D = \frac{\sqrt{\left(\langle D_s^2 \rangle - \langle D_s \rangle^2\right)}}{\langle D_s \rangle} = \sqrt{\frac{\kappa_2}{\kappa_1^2}} \qquad (1.53)$$

Assuming a certain particle size distribution function, for example Gaussian or Poisson, the size polydispersity,

$$\sigma_R = \frac{\sqrt{\left(\langle R_H^2 \rangle - \langle R_H \rangle^2\right)}}{\langle R_H \rangle} \qquad (1.54)$$

can be calculated from the polydispersity of the diffusion coefficients σ_D.

It has to be taken into account as already mentioned that for polydisperse samples the average selfdiffusion coefficient determined from the correlation function $F_s(q,\tau)$, e.g., by Cumulant analysis, is q-dependent. Therefore, it is also called apparent diffusion coefficient $D_{app}(q)$. This quantity is defined as:

$$D_{app}(q) = \frac{\sum n_i \cdot M_i^2 \cdot P_i(q) \cdot D_i}{\sum n_i \cdot M_i^2 \cdot P_i(q)} \qquad (1.55)$$

with n_i the number density of scattering particles of species i, M_i their particle mass, $P_i(q)$ the particle form factor and D_i the corresponding selfdiffusion coefficient. In Eq. 1.55, for simplification contributions of nondiffusional relaxation processes, which have to be considered for nonspherical particles, are ignored.

The "true" average diffusion coefficient $\langle D_s \rangle_z$, which (like the radius of gyration, see Eq. 1.23) above) is a z-average, is determined by extrapolation of the apparent diffusion coefficient towards zero scattering vector q, since in this limit $P_i(q) = 1$ for all particle species, and also nondiffusional processes like rotation or polymer segment fluctuations do not any longer contribute to the correlation function. For small particles (10 nm <particle radius R_H <100 nm), this extrapolation in analogy to the Zimm equation (see static light scattering) is given as:

$$D_{app}(q) = \langle D_s \rangle_z \left(1 + K \langle R_g^2 \rangle_z q^2\right) \tag{1.56}$$

The constant K depends both on sample polydispersity and on the particle topology (sphere, cylinder, etc.). Only for samples consisting of monodisperse spheres, $K = 0$ and $D_{app}(q) = D_s$. As a consequence of Eq. 1.56, plotting $D_{app}(q)$ vs. q^2 in experimental practice may result in a linearly increasing function, whose intercept with the $q = 0$ axis yields the z-average diffusion coefficient $\langle D_s \rangle_z$ (and therefore an inverse z-average hydrodynamic radius $\langle R_H^{-1} \rangle_z$).

The reason for the linear increase of $D_{app}(q)$ with increasing q due to polydispersity effects in case of spherical particles, where only translational diffusion contributes to the correlation function, is simple to explain: the static scattering intensity $I(q)$, originating from a given scattering particle species of size R_H, determines its respective contribution to the correlation function $F_s(q,\tau) = \int P(D_s) \exp(-q^2 D_s \tau) dD_s$. As shown above, $I(q)$ depends both on particle mass concentration and particle form factor (see Eq. 1.20). Whereas obviously the concentration is not depending on the scattering vector q, the particle form factor $P(q)$ in case of particles larger than 20 nm certainly is. For illustration, we next consider the particle form factors of 3 spherical scattering particles of different sizes as shown in Fig. 1.14.

With increasing q, $P(q)$ of the larger particles decays first. For this reason, the relative contribution of larger particles to the correlation function measured for

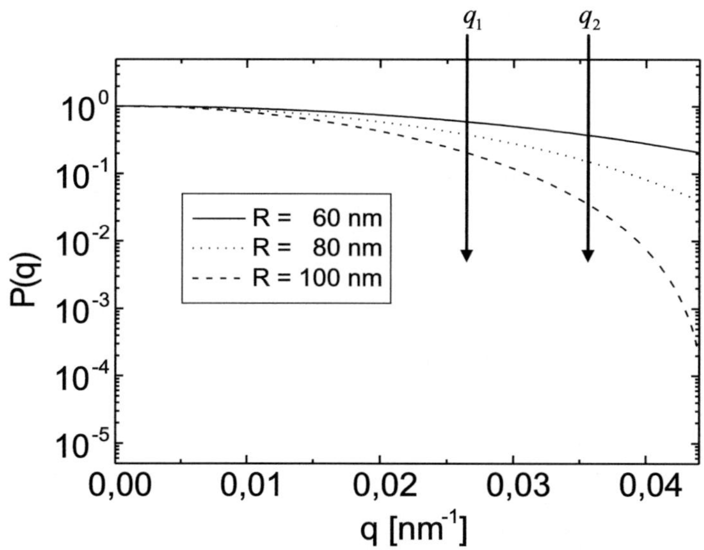

Fig. 1.14. Reason for the increase in D_{app} vs. q^2 for polydisperse systems: shown is the q-dependence of the particle form factors for spherical particles of three different sizes.

a polydisperse sample in a dynamic light scattering experiment also decreases with increasing q. This leads to an increasing contribution of the smaller particles to the dynamic light scattering signal with increasing q, and correspondingly must lead to an apparent increase of the average diffusion coefficient.

The ρ-Ratio

At the end of this brief review on the theoretical background of light scattering, I should not forget to mention the so-called ρ-ratio, an experimental quantity derived from combining the particle size characteristics determined from static and dynamic light scattering measurements. The ρ-ratio provides an important indication of the scattering particle topology especially for comparatively small particles (size 10–100 nm), where a detailed analysis of the particle form factor (see Eqs. 1.25–1.28) due to the limited length scale of the light scattering experiment is not possible. It is simply defined as:

$$\rho = \frac{R_g}{R_H} \tag{1.57}$$

Theoretically calculated values of ρ-ratios for the most important particle topologies have been summarized in Table 1.3.

Table 1.3. ρ-ratio for the most-typical particle morphologies

Topology	ρ-ratio
Homogeneous sphere	0.775
Hollow sphere	1
Ellipsoid	0.775 - 4
Random polymer coil	1.505
Cylinder of length l, diameter D	$\frac{1}{\sqrt{3}} \cdot \ln\left(\frac{l}{D} - 0.5\right)$

2 Experimental Setups

In this chapter, the experimental light scattering techniques most commonly used today will be reviewed briefly. Here, it has to be pointed out that the development of new technical approaches to light scattering is an ongoing process. Therefore, the reader should be aware that some very interesting and, in respect to new applications, important recent light scattering setups may exist that are not considered in this book.

In Chap. 2.1, the single angle light scattering setup based on a goniometer to vary the scattering angle will be described. Its disadvantage both for dynamic and static light scattering experiments is the long time needed for a single accurate angular dependent measurement. Chapter 2.2 presents the more recent technical advances allowing the simultaneous measurement of scattered light intensity at several scattering angles, and thereby reducing the overall measurement time. Such simultaneous measurements can cover the whole q-range at once in contrast to the conventional goniometer-based technique, where each scattering angle is measured separately in a sequence. As an alternative to conventional single angle light scattering detectors, a CCD chip with a lens setup in front can be positioned at one scattering angle and used as an array detector to observe a 2D image of the scattered intensity. Interferences of light scattered from different particles undergoing Brownian motion cause a pattern of bright and dark spots fluctuating with time. The bright spots here correspond to constructive interferences of slightly different scattering vector, and one of these fluctuating spots is called coherence area or speckle. With the help of the CCD chip as an array detector, several of these speckles can be monitored simultaneously at nearly identical scattering vector (scattering angle uncertainty less than $\pm 0.5°$). This approach allows partial replacement of the time averaging, needed to determine the autocorrelation function in dynamic light scattering experiments, by ensemble averaging, and has successfully been employed to study very slow diffusional processes of colloidal particles in highly viscous solvents or concentrated colloidal systems. In Chaps. 2.3 and 2.4, I will also describe some recent technical developments suitable to characterize optically nontransparent samples, where multiple scattering leads to erroneous results if standard light scattering techniques are employed.

2.1 Single Angle Scattering Using Goniometer Setups

All standard single angle light scattering setups, commercial or home-built, consist of the following components:

1. The incident light source, typically a laser (for example gas ion, HeNe, solid state or, nowadays, even laser diodes)
2. The light scattering cell, in most cases a cylindrical quartz glass cuvette of outer diameter between 10 and 30 mm, embedded, if possible, within an index matching and thermostating bath.
3. The detector, either a photo multiplier tube or the more recently available, very sensitive avalanche photo diode (APD), and its associated optics (pinhole or optical fiber, see below), mounted on the arm of a goniometer.
4. The electronic hardware components associated with the detector used for signal processing (computer, hardware correlator, etc.).

This setup is shown in Fig. 2.1.

Let us consider the components of the goniometer-based single angle light scattering setup in more detail:

1. The light source, in many cases still a continuous gas ion laser (typically Ar^+ or Kr^+), provides coherent and monochromatic light of power between a few milliwatts (mW) and several watts (W). In practice, the light intensity needed for a successful scattering experiment depends on the sensitivity of the optical detector, and on the scattering power of the sample itself as determined by size, concentration, and refractive index increment of the solute particles (see Chap. 1). Some setups use solid state lasers, which have been improved technically concerning their light quality (coherence, stability) in the last 5–10 years. These solid state lasers are much easier to handle than the gas lasers, since they are much smaller and less heavy and, most important, afford no external water cooling circuit. With the recent development of very sensitive light scattering detectors like the avalanche photo diode (APD), weak HeNe lasers (power 22 mW) become more frequently used in light scattering experiments due to the excellent optical properties of the emitted light and their simple handling. Some modern compact instruments employ small laser diodes, which nowadays are available with highly stable and coherent light emission as well as high laser power (50 mW and higher).

As shown in Fig. 2.1, typically the primary laser beam is guided and focused onto the sample by optical mirrors and lenses. The laser beam diameter within the sample, adjusted in this way by optical components, is well below 1 mm, which defines the scattering volume (see Fig. 1.5). Importantly, the laser emits light of a certain polarization (= direction of the electric field vector of the emitted light). Since this polarization determines the scattered intensity, it has to be specified: typically, vertically polarized light is used, meaning the electric

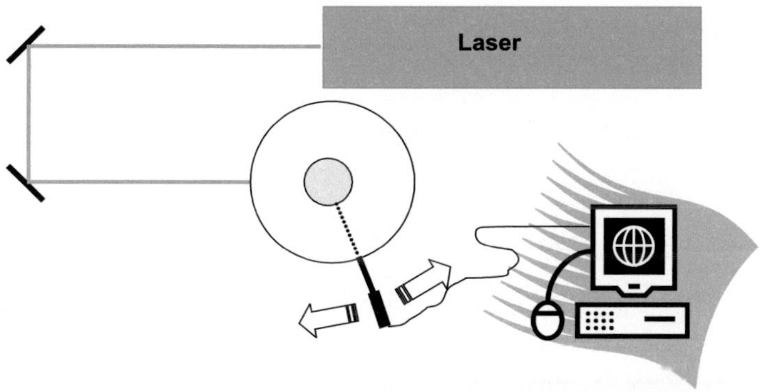

Fig. 2.1. Standard single angle light scattering setup (*top view*).

field vector is perpendicular to the scattering plane defined by incident laser beam and position of the optical detector with respect to the sample. In this case, the scattered intensity detected from an optically isotropic sample (either a pure solvent or a solution of very small (size <20 nm) particles) and normalized by the scattering volume, as described in Chap. 1, should be independent of the scattering angle, whereas horizontal polarization leads to a minimum in scattered intensity at 90°. This provides the means of adjusting the laser to the usually desired vertical polarization: using an optically isotropic sample, the polarization direction of the incident laser beam is tuned with either a half-wave plate or a polarization filter. Proper adjustment of vertical polarization then is identified as the optical alignment where the scattered intensity detected at 90° and normalized by the scattering volume (see Fig. 1.5) assumes its maximum value.

2. Using an index matching bath around the cylindrical light scattering cuvette is important to suppress unwanted diffraction of the incident and the scattered light at the sample-air-interfaces. Such diffraction could significantly change the actual scattering vector, thereby leading to systematic errors in the detected angular-dependent scattering intensity $I(q,t)$.

3. The detector optics determines the horizontal and vertical dimensions of the scattering volume, whereas its depth is defined by the width of the incident laser beam. Band pass filters with high transmission at the wavelength of the incident laser light are often used in front of the detector to suppress undesired contributions of stray light or fluorescence from the sample to the detected intensity. Additionally, some experiments, for example, detection of rotational diffusion by dynamic light scattering, need a polarization filter which, in this case, is called an analyzer, in front of the detector.

In the experimental practice of dynamic light scattering, several coherence areas or speckles are detected simultaneously. Therefore, the intercept of the normalized intensity correlation function, also called coherence factor, deviates from the theoretically expected value 1.0:

$$f_c = \frac{< I(q,t)^2 >}{\langle I(q,t)\rangle^2} - 1 < 1 \tag{2.1}$$

Accordingly, the Siegert relation (see Eq. 1.50) has to be modified:

$$g_2(q,\tau) = 1 + f_c \cdot g_1(q,\tau)^2 \tag{2.2}$$

The smaller the scattering volume defined by the detector optics, the lower the number of speckles, and correspondingly the larger the coherence factor f_c. On the other hand, a smaller scattering volume leads to a decrease in the overall scattered intensity and therefore to an increase in the signal-to-noise-ratio. In practice, sometimes one has to compromise between these two effects: in many dynamic light scattering experiments, especially in case of older instrumentation where the optical detectors are less sensitive, coherence factors in the range $0.3 < f_c < 0.6$ are used. The scattering volume and the corresponding f_c are adjusted either by pinhole setups or, more recently, by optical monomode or multimode fibers. A detailed experimental comparison of the two detector setups (pinhole and optical fibers) has been presented by Vanhoudt and Clauwaert [2.1]. The authors used a bimodal spherical colloid suspension as a testing sample. I will review their experiments and data analysis in more detail in Chap. 5.1, since it is a very illustrative example for current state-of-the-art performance and data analysis of dynamic light scattering experiments. Concerning the experimental detector setup itself, the authors conclude that "the best choice for an optical receiver in a light scattering setup which is supposed to be used for both SLS and DLS experiments is still a classical pinhole receiver with an experimental coherence factor between 0.4" and 0.7"." The largest disadvantage of this type of receiver is impracticable handling due to its size and weight. On the other hand, fiber receivers according to Vanhoudt and Clauwaert are more difficult to align for optimum detection efficiency and are not recommended for single angle SLS experiments, whereas they work very well for experimental setups designed only for single angle DLS experiments. Here, a single mode fiber detector is the best choice. For simultaneous multiangle SLS and DLS experiments (see Chap. 2.2), the authors recommend the use of few-mode fiber receivers. It should be noted that optical fibers as well as light scattering detectors have been technically improved since ref. [2.1] was published in 1999, and nowadays fiber detectors are recommended for any light scattering experiment due to their compact handling and comparatively simple optical alignment.

In a standard single angle scattering setup, detector optics and detector are mounted on the arm of a goniometer. The typical distance between optical detector and sample lies between 10 cm and 50 cm. The position of the detector is changed by a step motor in an angular range of typically 20° to 150° with step size 5°–10°. This angular range is limited by the primary laser beam and the transmitted laser beam, whereas the step size is limited by the total measurement time.

Especially at very small scattering angles, the transmitted primary beam may interfere with the detected scattered intensity and therefore alternative experimental setups are needed. Examples will be presented in Chaps. 2.2 and 5.3.

4. The scattered intensity $I(q,t)$, which is monitored by the optical detector, is typically digitized and stored in a computer. These raw data have to be further processed: in case of static light scattering, particle form factor models can be fitted to the data, as will be shown for some exemplary cases below. In a dynamic light scattering experiment, the raw data have to be converted to the time intensity correlation function $g_2(q,\tau)$: this is done online during the measurement using a very fast so-called hardware correlator, like for example the ALV 5000 from ALV, Langen, Germany. Alternatively, a large data set $I(q,t)$ is stored on the hard disk of the computer, and later the correlation function $g_1(q,\tau)$ is calculated according to Eqs. 1.49 and 2.2, using a self-written software algorithm. Since, at least for present-day computers, this software approach is not as effective as the specially designed hardware correlators, the data have to be stored first and processed separately. However, the latter approach is more flexible if one wants to calculate unusual correlation functions, for example, cross correlations of multiple speckles. More important, it is the only possibility for CCD-chip-based dynamic light scattering, where $I(q,t)$ from up to several thousand speckles can be monitored simultaneously, and correspondingly several thousand correlation functions can be calculated in parallel (see Chap. 2.2).

Finally, it should be noted that the whole light scattering setup illustrated in Fig. 2.1 has to be placed on a special optical table to isolate it from external vibrations. Such disturbances might not only ruin the optical alignment of the laser, sample, and detector itself, but also may create unwanted contributions to the correlation function measured in the dynamic light scattering experiment.

2.2 Simultaneous Multiangle Scattering

There exist several advantages of the simultaneous detection of the scattered light at multiple scattering angles compared to the single angle scattering setup illustrated in Fig. 2.1.

1. First, let us consider the so-called multiangle laser light scattering, briefly called MALLS or MALS. The typical MALS-setup is illustrated in Fig. 2.2. The first MALS-setups were home-built at university laboratories, but MALS recently also has become commercially available due to its importance as a fast and absolute particle characterization method in combination with analytical separation techniques, such as gel permeation chromatography (GPC) or field flow fractionation (FFF). Whereas the separation method separates a polydisperse sample into rather monodisperse fractions, the light scattering detector here is used to

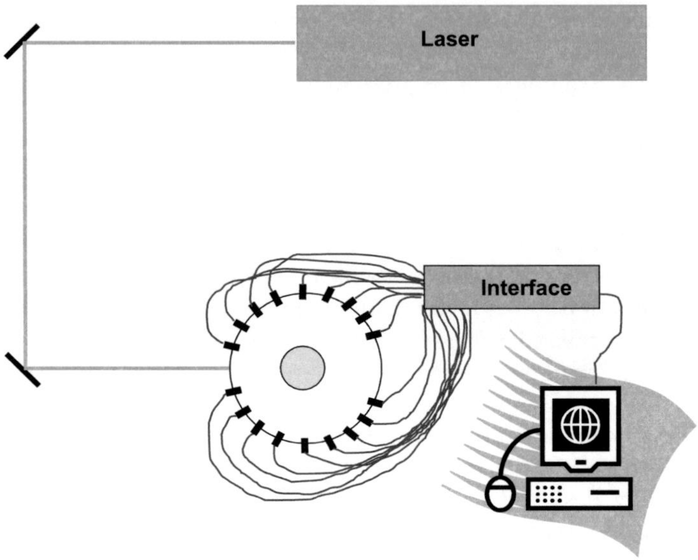

Fig. 2.2. Standard MALS setup (*top view*).

provide the sample characteristics of these fractions, such as molar mass, radius of gyration and hydrodynamic radius. The fractions pass a flow-through-cell which is embedded in an index-matching bath and illuminated by a laser. During the short period of time a given fraction migrates through this cell, $I(q,t)$ is monitored simultaneously at several (typically 20–40) fixed angular positions concentrically surrounding the sample cell. A set of optical fibers is used to collect the light scattered intensity at a given scattering angle, and transfers it to the electronic components used for signal digitization and processing. Nowadays, MALS setups are suitable both for SLS and DLS measurements, although due to difficulties in alignment and worse signal-noise-ratio of MALS compared to single angle scattering setups, the particle characteristics in most cases are not as accurate and reliable as those determined with a standard goniometer-based light scattering setup (Fig. 2.1). Commercially available MALS-setups will be presented in Chap. 3, the description of home-built setups can be found in the literature. One experimental example illustrating the potential of GPC-MALS will be reviewed in Chap. 5.2.

2. Another way of detecting light scattering intensities simultaneously at various scattering angles is using the CCD-chip of an optical camera as area detector. This approach has been used successfully in two fundamentally different ways:

(i) One application of a CCD area detector is to measure light scattering at very small scattering angles usually not accessible by the goniometer setups, which are typically limited to scattering angles $\theta > 20°$. To detect the scattered light at such

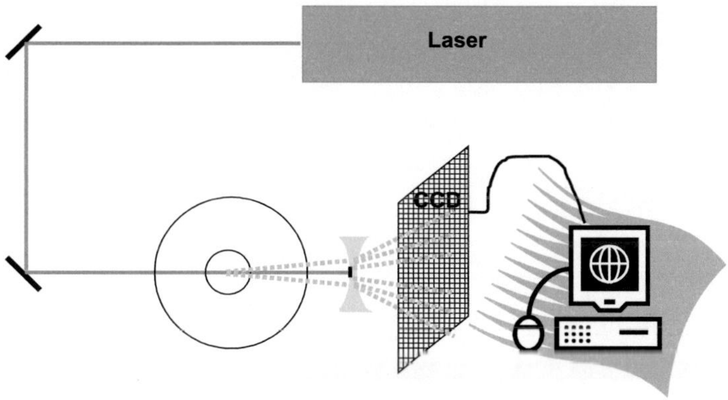

Fig. 2.3. Setup for small-angle light scattering based on CCD array detectors (*top view*, not on-scale!).

very small scattering angles, the CCD chip is positioned in direction of the transmitted laser beam behind the sample cell. The primary beam itself is blocked, using a so-called beam stop, and the small-angle scattering speckles are focused onto the CCD chip using an appropriate setup of optical lenses. In this way, even scattering angles down to 1° can be detected. For isotropically scattering samples, a concentric array of virtual optical detectors can be defined around the position of the blocked transmitted laser beam: for this purpose, images of the speckle pattern detected by the CCD chip are digitized at defined time intervals, using a frame grabber board (a 2D-digitizer), and stored on a computer. A digitized picture contains typically several hundred times several hundred picture elements (pixels). Since the CCD chip serves as an array of optical detectors, one can define a small area of a given image, typically 2 x 2 or 3 x 3 pixels^2, as a single virtual detector. The stored image data provide access to a whole set of detected scattered intensity time traces $I_n(q,t)$, where n is the index of one of potentially several thousand of these virtual detectors within the image. Concentric circles of these virtual detectors surround the blocked primary beam. Importantly, in case of isotropic samples the detectors on a given circle all correspond to an identical scattering vector q. Therefore, signals from these virtual detectors can be ensemble averaged to improve the statistical accuracy of the measured correlation functions. This approach was first used by Wiltzius et al. [2.2]. Their setup, which can be used both for static and dynamic small angle light scattering, is shown in Fig. 2.3 and will be discussed in more detail in Chap. 5.3.
Convex lenses or, as shown in Fig. 2.3, concave lenses can be used to define the scattering vector regime visible on the CCD chip, and either enhance the scattering angle range towards larger scattering angles (concave), or increase the angular resolution at very small scattering angles (convex).

(ii) The second application of a CCD camera chip as detector array leads to an enhancement of single-angle dynamic light scattering by replacing time averaging by ensemble averaging, and thereby shortening the overall measurement

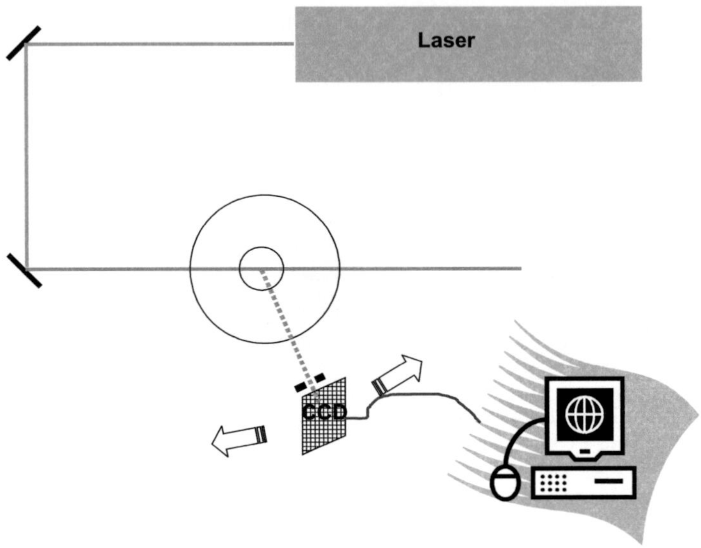

Fig. 2.4. Setup for multi speckle correlation spectroscopy (MSCS) (*top view*, not on-scale!).

time. This approach provides experimental access to very slow diffusion in dynamic light scattering. Here, the CCD chip is positioned at a given scattering angle with a pinhole setup in front, and used as a detector array to monitor several "speckles" or coherence areas simultaneously at nearly identical q (scattering angle uncertainty less than +/–0.2°). The corresponding experimental setup is shown in Fig. 2.4.

As already described above, a time series of images is digitized and stored on the computer (typically 50,000 images of 256 x 125 pixels^2 at time intervals $n \cdot \Delta t$, with $\Delta t = 0.2\,s$, corresponding to a total measurement time of 10000 s). After the measurement, each image is loaded from the hard disk, divided into a set of typically 100 statistically arranged virtual detectors of 2 x 2 pixels^2, and the time intensity correlation functions of the signals of these virtual detectors (which are the sums of the intensities of the four corresponding pixels, respectively) are calculated using a home-made software algorithm. These 100 individual time-averaged intensity autocorrelation functions are ensemble averaged to yield the time- and ensemble-averaged intensity correlation function $\langle g_2(q,\tau)\rangle$ in a correlation time regime $0.2\,s < \tau < 1000\,s$. Compared to a conventional DLS experiment, where the correlation function is determined from the detected fluctuating scattered intensity $I(q,t)$ by time averaging only, the additional ensemble averaging leads to a strong decrease in overall measurement time at a given statistical accuracy. In this case, also extremely slow dynamic processes become experimentally accessible by the DLS technique. More details of this approach as well as an experimental application will be reviewed in Chap. 5.3.

At the end of this chapter, I should not forget to mention the disadvantages of the otherwise so useful CCD camera-chip-based setups in comparison to conventional photomultiplier or photodiode systems. Since the signal $I(q,t)$ is measured by a combination of CCD chip, digital frame grabber and computer, a severe limitation of the technique is the shortest possible time interval Δt between successive images. For technical reasons, the minimum time interval Δt is in the order of 0.1 s compared to about 100 ns for the conventional dynamic light scattering setup. Although high speed cameras with $\Delta t \ll 1ms$ recently became available, frame grabbing and data transfer still take a lot of time. A second draw back of the method is the comparatively low dynamic range of the CCD camera chip itself, that is, its incapability to resolve scattered light intensities covering several orders of magnitude, which is no problem for the conventional detectors (photomultiplier and APD). In addition, digitization of the intensities using the frame grabber further limits this dynamic range. The first multispeckle correlation experiments (MSCS) with CCD detector (of the author of this book himself) have been conducted with an 8 bit frame grabber, limiting the dynamic range of detected scattered intensities $I(q,t)$ to roughly two decades (0–255) only. Although modern commercially available frame grabbers have a dynamic range of 12 or even 16 bit, they are still limited to a dynamic data range of four decades.

2.3 Fiber-Optic Quasielastic Light Scattering

A simple-to-use method to determine the particle size in turbid dispersions has been developed by Auweter and Horn of BASF-company: fiber-optic quasielastic light scattering, also called FOQELS. The experimental setup, shown in Fig. 2.5, has been described in detail in the paper by Wiese and Horn [2.3] which will be reviewed in Chap. 5.3.

A single-mode optical fiber is used both to irradiate the sample and to collect the scattered laser light, which then is guided to a photomultiplier (PM) connected to a computer equipped with a hardware correlator. The single-mode fiber selects scattered light that way so that only the part of the speckle pattern formed by backscattered light waves will be guided to the detector. This detection scheme therefore suppresses multiple scattering usually found in turbid solutions, and defines the scattering angle as 180°. The authors have successfully demonstrated that their technical setup can be used to measure the diffusion of latex particles in turbid dispersions. However, one should note that in case of highly concentrated dispersions not the single particle tracer diffusion coefficient but the so-called collective diffusion coefficient is contained in the correlation function $g_1(q,\tau)$. Depending on particle interactions, selfdiffusion and collective diffusion coefficient may strongly deviate from the value predicted by the Stokes–Einstein–equation (Eq. 1.47), and an accurate determination of the particle size in a concentrated turbid dispersion may become practically impossible.

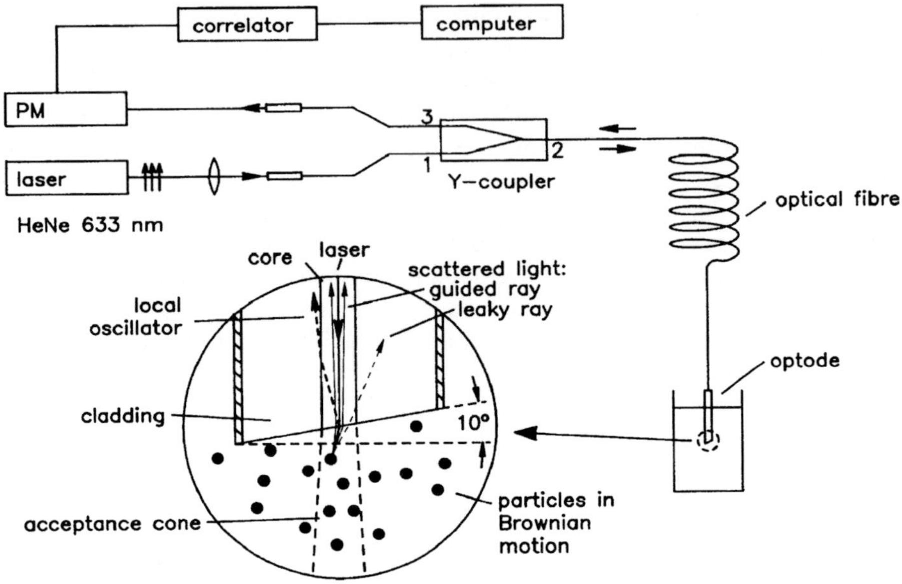

Fig. 2.5. Setup for fiber-optic quasielastic light scattering (FOQELS). Reused with permission from H. Wiese and D. Horn, Journal of Chemical Physics, 94, 6429 (1991), Copyright 1991, American Institute of Physics (ref. [2.3]).

2.4 Crosscorrelation Techniques – Dual Color and 3D Dynamic Light Scattering

An alternative technique to suppress multiple scattering in DLS experiments is the so-called dual color crosscorrelation. This method consists of a combination of two conventional dynamic light scattering setups, each employing laser light of a different wavelength. The principle of this approach is that multiple scattering contributions are discarded from the correlation function by cross-correlating the temporal fluctuations of two simultaneously performed geometrically different scattering experiments, which share the same scattering vector $\vec{q}$ and the identical scattering volume. The two independent experiments are realized in experimental practice by two laser beams with different wavelengths from the same light source, and two optical detectors each equipped with a narrow bandwidth filter in front. Typically, a multiline Ar^+-laser is employed, using the two main wave lengths $\lambda_1 = 488$ nm and $\lambda_2 = 514$ nm. One experimental example of this technique will be briefly reviewed in Chap. 5.3 of this book.

An alternative to the dual color crosscorrelation technique is the 3D setup. Here, the two simultaneous single-angle dynamic light scattering experiments at identical scattering vector and identical scattering volume are not realized by incident light beams of two different wavelengths, but by splitting the incident laser beam into two and positioning two detectors at identical scattering angles

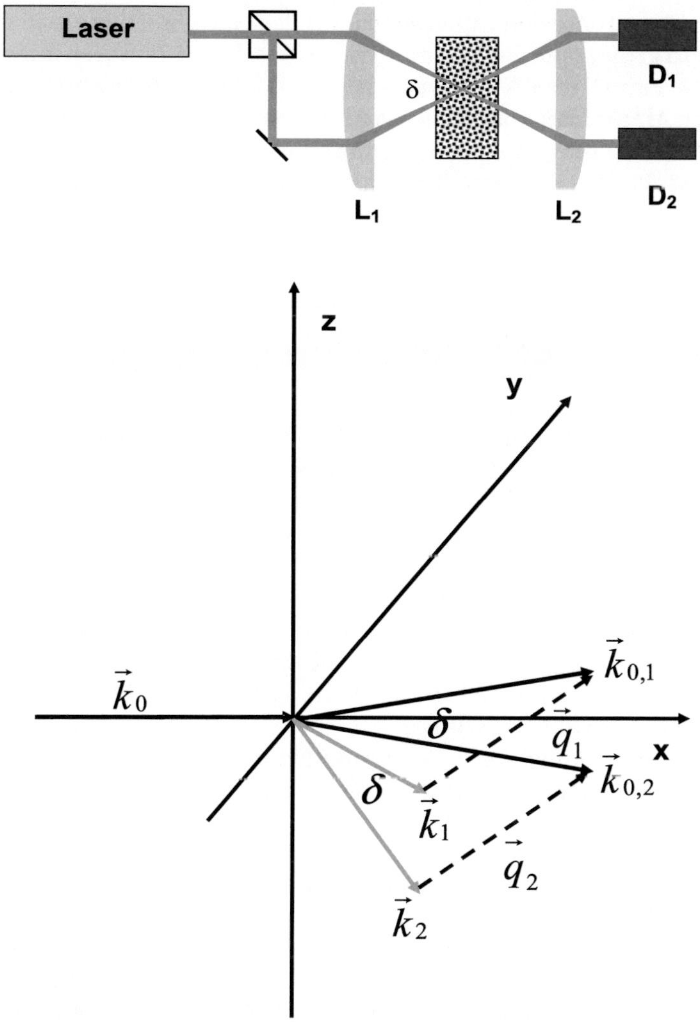

Fig. 2.6. Setup for 3D crosscorrelation dynamic light scattering (*top*) and corresponding scattering vectors (*bottom*).

slightly above and below the scattering plane. The corresponding setup and scattering geometry are shown in Fig. 2.6.

L_1, L_2 are lenses to focus the incident and the detected light within the scattering volume, D_1, D_2 are the optical detectors (monomode fiber plus photomultiplier) connected to the correlator hardware. Urban and Schurtenberger implemented a HeNe laser (Spectra Physics, model 127), operating at 633 nm, as light source to extend the measurable range of sample turbidity in respect to an Ar^+ ion laser used in their first experiments. The illuminating light beam is split into two parallel beams which are focused by a lens (planar convex, focal length 90 mm, diameter

63 mm, Spindler & Hoyer) onto the scattering cell. This lens was chosen in order to adjust the beam waist at the scattering volume to values less than 50 mm at an intersection angle $\delta = 18°$, which results in a suppression of double scattering by a factor of 4e-3 in case an Ar^+ ion laser is used and even lower in case of a HeNe laser. Let us consider the scattering geometry for this setup shown in Fig. 2.6 (bottom) in more detail: here, x is chosen as the direction of the primary laser beam before splitting, and the two incident laser beams are located in the xz-plane, intersecting in the scattering volume at an angle $\delta = 18°$. The scattered light wave vectors $\vec{k}_1, \vec{k}_2$ are rotated with respect to the incident light wave vectors $\vec{k}_{0,1}, \vec{k}_{0,2}$ by the scattering angle θ out of the xz-plane. This arrangement is symmetrical with respect to the xy-plane, leading to identical scattering vectors $\vec{q}_1$ and $\vec{q}_2$.

In case of turbid samples, the optical path length, and correspondingly the size of the sample cell, is very important to avoid a strong reduction of the single scattered light and the subsequent signal reduction. To adjust the optical path length to the small value of 1 mm, Urban and Schurtenberger used square cells of 10-mm thickness and positioned them such that the scattering volume is located in a corner of the cell. For experiments at scattering angles different from 90°, the cell was turned by half the rotation angle 90°- θ in order to recover a symmetrical situation in which the displacement of the incident and scattered beam due to diffraction at the cell wall almost cancel. For more details of the 3D crosscorrelation procedure and practical examples, the interested reader is referred to ref. [2.4] and to the web page of the company LS Instruments (http://www.lsinstruments.ch/3DDLS.htm).

Importantly, the 3D setup is much more robust concerning the stability of the experimental setup compared to the dual color crosscorrelation, where temperature fluctuations on the order of 0.01 °C already can be sufficient to spoil the optical alignment. Also, a 3D dynamic light scattering machine has recently been comercialized by Professor Schurtenberger and coworkers (http://www.lsinstruments.ch/3DDLS.htm). This method provides an excellent alternative to the also commercially available fiber optic light scattering technique (see Chap. 3.3) for the characterization of turbid samples by dynamic light scattering. The advantage of the 3D setup, compared to the FOQELS technique, is that the scattering angle can be varied, which is an important prerequisite especially for the characterization of polydisperse samples.

To conclude this section, it should be noted that other experimental techniques of light scattering, especially of dynamic light scattering, not discussed in this book have been developed in the last 10–20 years. One prominent example is diffusing wave spectroscopy (DWS), which is also very useful to characterize nontransparent samples. However, the theoretical principle of this technique is different from the approach illustrated in Chap. 1 of this book, and the method itself so far has rarely been used for practical particle characterization.

References

2.1 Vanhoudt J, Clauwaert J (1999) Experimental comparison of fiber receivers and a pinhole receiver for dynamic and static light scattering. Langmuir 15:44

2.2 Wong APY, Wiltzius P (1993) Dynamic light scattering with a CCD camera. Rev Sci Instrum 64:2547

2.3 Wiese H, Horn D (1991) Single-mode fibers in fiber-optic quasielastic light scattering: A study of the dynamics of concentrated latex dispersions. J Chem Phys 94:6129

2.4 Urban C, Schurtenberger P (1998) Characterization of turbid colloidal suspensions using light scattering techniques combined with cross-correlation methods. J Colloid Interface Sci 207:150

3 Commercial Light Scattering Instruments

In this chapter, a list of commercially available experimental light scattering setups is presented. Here, the reader should note that many researchers, especially those working in academia, tend to combine commercially available components, like optical equipment (lenses, pinholes, mirrors), optical table, laser, X-ray goniometer, and a computer equipped with hardware correlator board, to build their own light scattering setup. This procedure has the advantage that the operator is not restricted to a "black box" but knows all aspects of data acquisition and data processing of the setup, which is sometimes necessary for people investigating new unknown structures by light scattering. Moreover, a home-built setup usually is more flexible and can therefore easily be adjusted to new academic problems. On the other hand, for routine nanoparticle or polymer characterization, where the particle topology itself is known in principle and only measurement of the unknown particle size is required, commercial light scattering machines, often also called particle sizers, definitely yield reliable data.

Nowadays, light scattering detectors are not only used as stand-alone characterization equipment, but frequently in combination with fractionation methods as gel permeation chromatography or field flow fractionation. This approach has become necessary since synthetic nanoparticles often show a pronounced polydispersity or even multimodal size distribution, and characterization of these polydisperse samples by light scattering (without previous sample fractionation) doesn't always provide reliable results. Fractionation of the polydisperse sample and simultaneous quantitative characterization of the better-defined, meaning less polydisperse, sample fractions by light scattering here is the proper solution. In this context, as mentioned above, simultaneous multiangle light scattering detectors, which provide rapid access to the angular dependence of the scattered intensity, are becoming increasingly important.

The commercial equipment listed in the following sections has been arranged following the order of principle setups presented in Chap. 2. The list endeavors to provide an up-to-date list of the best-known and, due to their long tradition, the most experienced companies in the field of light scattering, but does not claim to be complete. Certainly, some very recent technical developments as well as products from newer and/or smaller companies might have been overlooked by the author of this book, for which he would like to apologize at this point.

3.1 Single-Angle Light Scattering

Most of the following light scattering detectors are not based on a goniometer as shown in Fig. 2.1 and therefore do not allow flexible adjustment of the scattering angle. Instead, they operate at one fixed single scattering angle. For well-defined monodisperse systems, dynamic light scattering in this case yields reliable and accurate results. For polydisperse samples, like most synthetic nanoparticles or polymers, dynamic light scattering using various scattering angles as provided either by a multiangle setup (see Chap. 3.2.) or a goniometer setup (see, for example, Brookhaven BI-200SM, ALV /CGS, or home-built) is recommended: with such a dynamic light scattering equipment, the z-average inverse particle size $\langle R_H^{-1} \rangle_z$ can be and has to be determined by extrapolation of the apparent diffusion coefficients measured at finite scattering angles to zero scattering angle (see Chap. 1.3.).

A) Light scattering setups from the company ALV
(see http://www.alvgmbh.com/About_ALV/about_alv.html for more detailed information):
ALV / CGS-8F, a goniometer platform for integration of customer supplied lasers (e. g., Argon-Ion or frequency-doubled NdYAG), equipped with the ALV-Static & Dynamic Enhancer fiber optical detector.

B) Light scattering setups from the company Wyatt
(see http://www.wyatt.com/solutions/hardware/ for more detailed information):
WyattQELS (quasi elastic light scattering), a classical dynamic light scattering machine operating at one scattering angle only.
Dynapro Titan, a small volume, batch dynamic light scattering instrument to measure the hydrodynamic radius of nanoparticles, proteins, vesicles, viruses, colloids, etc., in solution.

C) Light scattering setups from the company Brookhaven
(see http://www.brookhaven.co.uk/products.html for more detailed information):
90Plus, which determines submicron particle size distributions for particles ranging from around 2 nm up to 5 μm by dynamic light scattering at two different scattering angles 90° or 15°.
BI-200SM, a research goniometer with wide and flexible angular range for both static and dynamic light scattering applications in particle and macromolecular characterization. The instrument accommodates a wide range of lasers and accessories.

D) Light scattering setups from the company Viscotek
(see http://www.viscotek.com/productprofiles.php4 for more detailed information):
Viscotek DLS, a state-of-the-art dynamic light scattering incorporating the advantageous Single Mode Fiber technology, and requiring only a small sample volume (10 μl) which is recoverable.

E) Light scattering setups from the company PolymerLabs
(see http://www.polymerlabs.com/pdi/products.htm for more detailed information):
PD2020 Detector, an ambient temperature static (Rayleigh) light scattering detector, operating at sample temperatures up to 50°C with a detector at both 90° and 15° to the incident laser beam. The instrument can be used to determine average molar mass, radius of gyration and other parameters for a wide range of molecular sizes, typically from a few 100 s of Daltons to 10,000,000 Daltons.
PDDLS/Cool Batch Sizing System, a dynamic light scattering-based molecular size analysis system which provides molecular sizing data (hydrodynamic radius) for macromolecules and nanoparticles in solution, ranging from 1.0 to 1000 nm at precisely controlled sample temperatures of 4° to 40°C.
PDDLS/Flow Laser Light Scattering Detector, a flow-mode dynamic light scattering detector (DLS) used in gel permeation chromatography (GPC) applications, alone or with the PD2020 static light scattering detector. The combination provides accurate molecular weight determinations and size in "real-time" in the same sample cell. This instrument can determine the hydrodynamic radius of molecules over a very wide range (1.5 nm to 1,000 nm).

3.2 Multiangle Light Scattering (MALS)

A) Light scattering setups from the company ALV
(see http://www.alvgmbh.com/About_ALV/about_alv.html for more detailed information):
ALV / CGS-8F, the goniometer platform for integration of customer supplied lasers (see Chap 3.1), fully equipped with parallel detection at eight angles with 17° angular displacement to each other. The detection channels are based on the proprietary ALV-Static & Dynamic Enhancer fiber optical detector and seven ALV-SIPC II single photon detectors plus one ALV / SO-SIPD "pseudo-cross-correlation".

B) Light scattering setups from the company Wyatt
(see http://www.wyatt.com/solutions/hardware/ for more detailed information):
DAWN HELEOS, a research-oriented multiangle static light scattering (MALS) instrument for absolute molecular weight, size, and conformation of macromolecules and nanoparticles in solution.
miniDAWN, a triple-angle static light scattering (MALS) detector for absolute molecular weight, size, and conformation of macromolecules in solution.

C) Light scattering setups from the company Brookhaven
(see http://www.brookhaven.co.uk/products.html for more detailed information):
BI-MwA, a multiangle fiber-optic system for characterizing macromolecules in solution by static light scattering.

D) Multiangle dynamic light scattering particle sizer from the company Beckmann Coulter (see http://www.beckmancoulter.com/products for more detailed information)

The N5 Submicron Particle Size Analyzer, a commercial dynamic light scattering particle sizing system using six different fixed scattering angles. The light source is a 25 mW HeNe laser, fixed scattering angles are 14.9°, 20.6°, 30.4°, 40.2°, 50.4°, and 90° addressable by the user with an auto-stepping motor routine for sequential angle analysis. The particle size range of this device ranges from 0.003 μm to 3 μm. Size distributions are calculated from the DLS data using the CONTIN program (see Chap 5.1).

E) Multiangle static light scattering particle sizer from the company Malvern (see http://www.malvern.com/ for more detailed information):

The Mastersizer 2000, a modular instrument designed for the measurement of the particle size distribution of wet and dry samples. It has a wide range of sample dispersion units including an autosampler. Due to its range of detector angles covering forward scattering, wide angle forward scattering, side scattering, and back scattering, and the use of two different wavelengths of incident light (red light: HeNe laser, blue light: solid state light source), the accessible particle size ranges from 0.02 μm to 2,000 μm.

3.3 Fiber Optic Quasielastic Light Scattering and 3D Crosscorrelation

A) FOQELS from the company Brookhaven (see http://www.brookhaven.co.uk/products.html for more detailed information):

FOQELS, a setup for noninvasive studies of particle size distribution and dynamics of submicron particles at high concentration using the fiber-optic backscatter technique.

B) 3D crosscorrelation dynamic light scattering setup from the company LS Instruments (see http://www.lsinstruments.ch/3DDLS.htm for more detailed information):

The 3D crosscorrelation instrument from LSI, using the 3D technology to simply filter single scattered light from the scattering signal and dismiss contributions from multiple scattering. Suitable for particle characterization even in case of turbid samples down to 5% optical transmission (in case of sub-mm path lengths). It is a monomode fiber goniometer system equipped with Flex correlator and a diode laser with wavelength of 680.4 nm (TUI Optics) as light source. The setup offers full software controlled operation at angles from 10°–150° and automated particle size determination (Cumulant fit and CONTIN Analysis, see Chap 5.1). In addition, its compact and robust design leads to easy alignment and maintenance.

4 Sample Preparation

Having studied the theory and experimental setup of the light scattering method in the preceding chapters, let us now consider how light scattering samples are prepared in practice.

The sample cells suitable for the experiment have to be chosen according to experimental conditions, such as light scattering detected offline or in combination with a fractionation technique, and sample parameters (amount, temperature, pressure, etc.). If light scattering detectors are used in combination with chromatography or field flow fractionation, special commercially available flow cells have to be used. For most discontinuous measurements of aqueous or nonaqueous solutions of the scattering nanoparticles or polymers, cylindrical quartz glass cuvettes with an outer diameter between 1 cm and 3 cm are useful (see Chap. 6 for commercial provider). For high temperature samples like nanoparticles dispersed in a polymer melt, special sample cells, often home-built, are needed. In this chapter about the preparation of light scattering samples, I will restrict myself to the most common experimental practice of light scattering: characterization of particles in solution at room temperature.

The sample parameters summarized in Table 4.1 are needed for quantitative light scattering data analysis and have to be determined either experimentally or to be looked-up in standard manuals like the well-known "Handbook of Physical Chemistry".

Note that there is a difference in the parameters needed for static light scattering or dynamic light scattering. The parameters listed in Table 4.1 follow from the key equations for static (Eqs. 1.9 and 1.21), Zimm analysis) and dynamic light

Table 4.1. Sample parameters needed for data evaluation in light scattering experiments

Parameter	Static light scattering	Dynamic light scattering
Solvent viscosity η	No	Yes
Solvent refractive index n_D	Yes	Yes
Sample concentration in g/L	Yes	No
Sample temperature	No	Yes
Refractive index increment dn_D/dc	Yes	No

scattering (Eq. 1.47, Stokes–Einstein–equation), respectively. The refractive index of the solvent is needed in all cases since it determines the magnitude of the scattering vector q according to Eq. 1.14. Whereas pure solvent quantities as viscosity η or refractive index n_D can be looked up in tables, for instance in the Handbook of Physical Chemistry, the refractive index increment dn_D/dc often has to be measured, for example, if the solute scattering particles consist of non-standard chemicals or mixed chemical components. For this purpose, commercial interferometers with sufficient accuracy are available from most companies who also provide commercial light scattering setups (see Chap. 3). In Chap. 6.3 of this book, solvent characteristics η and n_D as well as refractive index increments dn_D/dc for the most common solvents and polymer-solvent pairs used in light scattering experiments are presented.

4.1 Sample Concentration and Interparticle Interactions

Since we are interested in single particle characterization, the samples should be as dilute as possible to suppress interparticle interactions, which may create contributions from interparticle interferences (structure factor, see Chap. 1) to the angular dependence of the detected scattered intensity, and therefore spoil the form factor analysis in the static light scattering experiment. Also, the Stokes–Einstein–equation (Eq. 1.47) in this case is no longer valid, and diffusion coefficients determined by dynamic light scattering yield wrong hydrodynamic radii. On the other hand, the concentration of scattering particles within solution has to be large enough to create sufficient scattered intensity at all scattering angles of interest (typically in the range 30° to 150°). Here, the reader should remember that the scattered intensity depends on the following parameters: (i) mass or size of the scattering particles, which unfortunately cannot be adjusted to experimental needs; (ii) solute particle concentration, which should be as small as possible; and (iii) refractive index difference of solvent and solute particles or, more accurate, refractive index increment dn_D/dc, which can be adjusted by choosing an appropriate solvent. Importantly, one has to keep in mind that specific solvent-solute interactions, like the formation of aggregates or micelles or even particle disintegration, also have to be avoided.

Whereas nonpolar organic samples usually are not critical up to solute concentrations in the range 1 g/L to 10 g/L, charged systems, also called polyelectrolytes, are more difficult to handle both in water and in organic solvents due to the long-range Coulomb interactions between the scattering solute particles. These interactions may influence the experimental results both in static and dynamic light scattering measurement even at seemingly low particle concentrations <1 g/L. One common solution to this problem is to screen the disturbing Coulomb interactions by adding salt, while not changing the refractive index too much: typical salt concentrations vary from 10e-5 mol/L to 10e-2 mol/L depending on the sample concentration. For charged linear polymer chains, it is commonly agreed that at a molar concentration ratio c(monomer)/c(salt) <1 Coulomb interactions become negligible (see ref. [4.1]). In case of branched polyelectrolytes, for example copolymer micelles with charged corona chains in aqueous solution, Förster et al. [4.2] have encountered in their light scattering experiments no specific polyelectrolyte effects (like a structure factor peak in static light scattering or a second slow relaxation process ("slow mode") in dynamic light scattering) at c(monomer) /c(salt) <25. This leads to the conclusion that for branched charged polymers a much lower amount of added salt is sufficient to screen the Coulomb interactions than in case of linear polyelectrolyte chains, a consequence of the lower effective charge per monomer in case of dense polyelectrolyte systems like micelles, brushes, colloidal spheres, dendrimers, and so on. For a detailed review about light scattering studies of polyelectrolytes as well as the properties of polyelectrolytes in solution in general, the interested reader should consult the comprehensive nice review article by Förster and Schmidt about "polyelectrolytes in solution" [4.1].

The added salts most frequently used to screen charge interactions in aqueous solution are KBr, NaBr, or NaCl, whereas for organic solvents either LiBr [tetrahydrofurane (THF), dimethylformamide (DMF), dimethylacetamide (DMA)], or $NBut_4Br$ [chloroform ($CHCl_3$), dichloromethylene (CH_2Cl_2), o-dichlorobenzene] are recommended. Figure 4.1 shows the effect of added salt (here: 1.02 g/L LiBr and 0.872 g/L hyperbranched polyelectrolyte dissolved in DMF) on static light scattering results.

Whereas in case of no added salt a structure factor peak is clearly visible in the average scattered intensity, this peak disappears if LiBr is added, and the linear slope of Kc/R plotted versus q^2 , expected for static light scattering in the Zimm regime, is obtained. Figure 4.1 also shows a Zimm plot of the same sample at five different concentrations, from which weight average molar mass and z-average radius of gyration of the polyelectrolyte nanoparticles could be determined.

The effect of a structure factor, caused by Coulomb interactions between charged polyelectrolyte chains, on both dynamic and static light scattering, has been explored in detail by Förster et al. [4.3]: for salt-free aqueous solutions of quaternized poly(2-vinylpyridine), the reciprocal static scattered intensity plotted vs. q^2 showed a negative slope at low polymer concentration, which according to the Zimm equation (Eq. 1.21) would correspond to a negative radius of gyration R_g . Interestingly, if the polymer concentration was increased, Förster et al. obtained a straight line as expected for a Zimm plot of small noninteracting polymer coils. In this case, however, the extracted radius of gyration was larger than even expected for a rigid rod. If salt (KBr) was added to a salt-free highly diluted (c = 0.3 g/L) polyelectrolyte solution, the negative slope of the inverse scattered intensity versus q^2 changed from negative to positive as expected. Importantly, the radius of gyration decreased if the added salt concentration was increased from 0.01 mol/L to 0.1 mol/L, which indicates a collapse of the polyelectrolyte coil at high added salt concentration. The studies by Förster et al. illustrate an important aspect of the static light scattering characterization of polyelectrolytes not discussed so far: not only does the occurrence of a structure factor peak interfere with the analysis of the particle form factor. Importantly, there is also the possibility of a change of the particle structure itself depending on the concentration of salt, which has to be added to screen the interparticle interactions.

If in experimental practice one is not sure about the appropriate sample and salt concentrations, it is recommended to conduct the light scattering measurements at

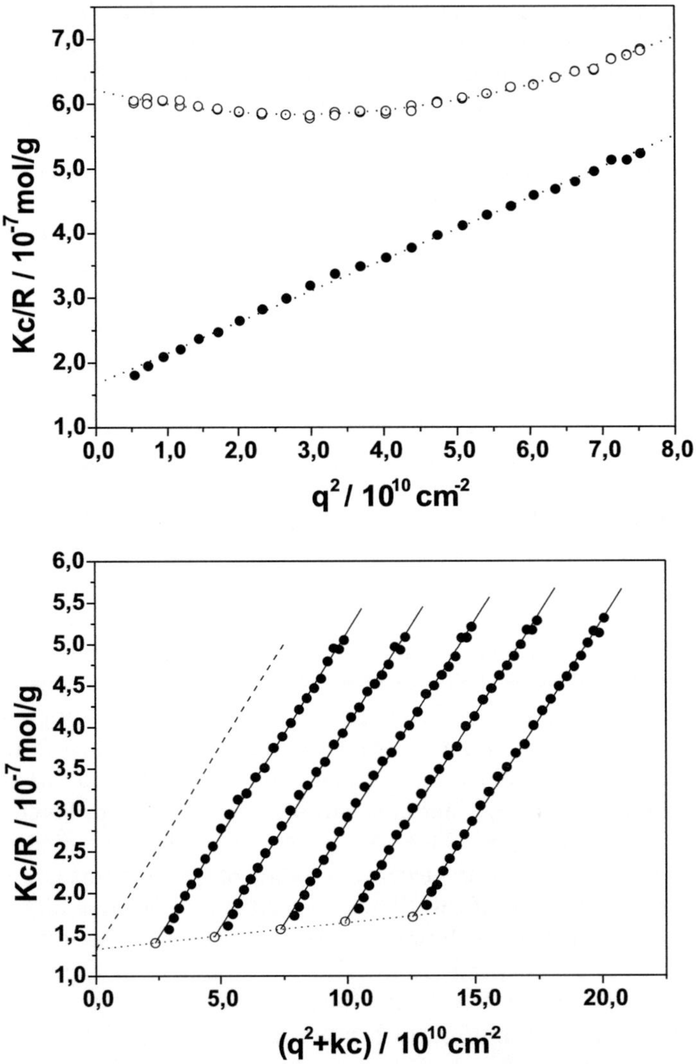

Fig. 4.1. Static light scattering results for charged supramolecular nanoparticles without (*open circles*) and with added salt (*filled circles*) (*top*), and Zimm plot for SLS from the noninteracting nanoparticles with added LiBr at various particle concentrations. [unpublished results, courtesy Dr. K. Fischer and Prof. A.D. Schlüter (Mainz, Zürich 2006)].

given salt concentration and various sample concentrations, covering at least one order of magnitude in the regime 0.1 <c(monomer)/c(salt) <1. If, in a dynamic light scattering experiment, the experimentally determined hydrodynamic radius of the scattering solute particles does not change with concentration, one can be fairly confident that the chosen concentration range is appropriate. For static light scattering, on the other hand, the shape of the scattered intensity plotted versus scattering vector q should not depend on concentration to ensure that only the particle form factor is measured and no unwanted side effects, like structure factor contributions, are detected (see Fig. 4.1).

The addition of added salt has an important consequence on the data analysis of the static light scattering experiment. Whereas for noncharged samples the refractive index increment dn_D/dc is needed to determine the scattering contrast, for charged particles also concentration fluctuations of the small ions (added salt and counter ions) contribute to the scattered intensity. As was shown by Vrij and Overbeek nearly 50 years ago, in this case the refractive index increments have to be measured under conditions defined by the Donnan equilibrium, that is not at constant salt concentration but at constant chemical potential [4.4].

Before we continue with the consequences of the addition of salt for the interpretation of static light scattering measurements, let us briefly review the Donnan equilibrium shown in Fig. 4.2.

As shown, the reader should imagine two chambers separated by a semipermeable membrane. Initially or before the Donnan equilibrium is established, the left chamber shall contain an aqueous solution of negatively charged nanoparticles and positive counterions only, whereas the right chamber contains an aqueous salt solution. For convenience, identical counter ions and cations of the salt have been chosen. The membrane is only permeable for the small ions (K^+, Cl^-), but impermeable to the larger negatively charged nanoparticles. To establish the electrochemical equilibrium, Cl^- ions will start to flow through the membrane from the right chamber into the left chamber, leading to an electric membrane

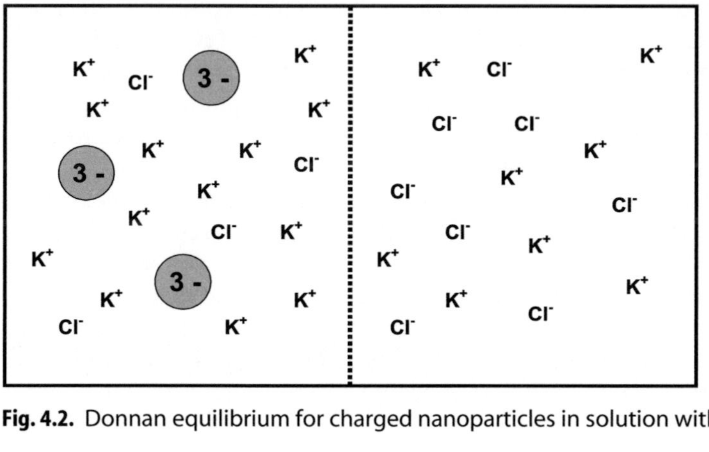

Fig. 4.2. Donnan equilibrium for charged nanoparticles in solution with added salt

potential attracting also the K^+ ions from right to left. The equilibrium situation illustrated in Fig. 4.2 is established, if the electrochemical potential of all small ions has become equal in the left and right chamber, respectively.

This equilibrium importantly corresponds to a difference in salt concentration between the two chambers: the left chamber containing the charged nanoparticles contains a lower amount of added salt than present in the right chamber, and a small electric membrane potential has built up between the two chambers. The difference in small ion concentrations also causes an osmotic pressure, leading to a solvent flow from the right chamber to the left chamber in case the chamber walls are not solid.

Coming back to static light scattering of charged particles in a solution containing added salt, Vrij and Overbeek showed that the apparent molar mass M_1* determined by Zimm analysis, using the concentration-dependent refractive index increments at constant salt concentration, is given as:

$$M_1* = M_1 \cdot \left[1 + \frac{(dn_D/dc_2)_{c_1}}{(dn_D/dc_1)_{c_2}} \left(\frac{dc_2}{dc_1} \right)_{\mu_2} \right]^2 \tag{4.1.}$$

Component 1 are the charged nanoparticles, component 2 is the added salt, and M_1 is the true molar mass of the nanoparticles. Note that the counter ions because of their usually small number (compared to the added salt ions) obviously are ignored in Eq. 4.1. The most obvious contribution to the term $(dc_2/dc_1)_{\mu_2}$ originates from the electrical double layer around the polyelectrolyte particles, which corresponds to a seat of negative adsorption of the small ions of the added salt. In simpler terms, the concentration of added salt will be smaller in the immediate vicinity of the nanoparticles than the solution average. Vrij and Overbeek tested their theoretical concept with turbidity measurements, which can be considered as a technically primitive version of modern static light scattering, for solutions of polymethacrylic acid (PMA) polyelectrolyte chains in aqueous HCl and in various salt solutions. Refractive index increments $(dn_D/dc_2)_{c_1=0}$ and $(dn_D/dc_1)_{c_2}$ were determined with a Rayleigh interferometer (fabricated by Zeiss). To measure the refractive index increment $(dn_D/dc_1)_{\mu_2}$, the solutions were dialyzed against a larger reservoir containing an aqueous salt solution with the same salt concentration c_2 (as used for the measurement of $(dn_D/dc_1)_{c_2}$) for 24 h, so the Donnan equilibrium was established. For typical salt species like NaCl or NaBr and salt concentrations $c_2 = 0.1$ Mol/L, the difference between $(dn_D/dc_1)_{c_2}$ and $(dn_D/dc_1)_{\mu_2}$ was the order of 10%. This leads to a corresponding difference in apparent molar mass M_1* and true particle molar mass M_1, with $M_1 > M_1*$. Since the scattered intensity depends on $(dn_D/dc_1)^2$ (see Eq. 1.10), the effect of the small ion concentration fluctuations on the molar mass detected by static light scattering is even much larger than 10%. We conclude therefore that in experimental practice the Donnan equilibrium should be established by dialysis of the light scattering sample without added salt versus a large reservoir of a pure salt solution (with concentration c_2). The large size of the salt solution reservoir, compared to the sample volume, guarantees that the salt concentration in the reservoir remains nearly constant while the equilibrium is established, since only a very small amount of salt will diffuse into the sample volume. Both refractive index increments and average scattered intensities have to be determined under these membrane equilibrium conditions to accurately determine the true particle molar mass M_1.

In case the static light scattering measurements were performed for aqueous sample solutions with different types of added salt, Eq. 4.1 provides the basis for determination of the true molar mass of the polyelectrolyte particles by conventional static light scattering, meaning even in case no dialysis is used to establish the Donnan equilibrium for measurement of $(dn_D/dc_1)_{\mu_2}$. By plotting $\sqrt{M_1{}^*}$ versus $M_2 \cdot (dn_D/dc_2)_{c_1=0}$, M_1 is obtained as the intercept with the axis $M_2 \cdot (dn_D/dc_2)_{c_1=0} = 0$.

4.2 Sample Purification

Having prepared a suitable sample solution, especially in case of weakly scattering small solute particles, it is important to remove any components not wanted to contribute to the scattered intensity, such as dust particles or air bubbles. Air bubbles in principle could be removed by careful ultrasonication of the sample, but it is more recommended to degas the solvents used for sample preparation. Dust is removed either by filtration or, if this is not possible, by centrifugation. In addition, the light scattering cell itself has to be cleaned from dust. This can be done, for example, by repeatedly flushing it with filtered acetone and then leaving it to dry under a dust-free special hood. A better method is to clean the cuvettes and the cuvette Teflon lids (!) with refluxing and therefore guaranteed clean acetone for 15–30 min (using a home-built distillation column like the one described in the famous book on physical chemistry of macromolecules by Tanford [4.5]), and then leaving them to dry in a dust free hood. To avoid dust contamination during this step, the sample cuvettes should be placed upside-down while drying.

Next, the sample solution itself is filtered, if possible, into these clean and dry cuvettes, using a handheld syringe and a porous membrane filter cartridge attached to the outlet of the syringe. The material of the filter (Teflon, cellulose, etc.) has to be chosen according to the solvent used for the light scattering sample, the pore size (typically 0.2–0.45 μm) according to the size of the scattering particles. One of the best-known commercial providers for a variety of suitable filters is the company Millipore (see Chap. 6). Importantly, the concentration of the sample may be changed by the filtration process due to adsorption of solute particles onto the filter membrane. An indication of strong adsorption of the scattering particles to the filter membrane is clogging of the filter, which should be avoided either by switching to a different membrane type or to a larger pore size up to 0.8 μm. To minimize the filtration losses in case of the absence of clogging, the filters have to be "conditioned" by filtering and discarding 1–2 mL of the sample solution before filling the light scattering cuvette. Still, filtration losses cannot be excluded after "conditioning" in every case. Whereas they play no role for dynamic light scattering experiments (see Table 4.1), it is crucial to detect them, and therefore accurately know the actual sample concentration, for static light scattering experiments. Here, the only possibility is to quantify the solute particle concentration before and after filtration, using, for example, uv/vis absorption spectroscopy. Only in case filtration does not work for a given sample, for example, due to serious filtration losses, centrifugation of dust particles onto the sample cuvette bottom and careful handling of the sample cell thereafter can provide a solution.

References

4.1 Förster S, Schmidt M (1995) Polyelectrolytes in solution. Adv Polym Sci 120:51
4.2 Förster S, Hermsdorf N, Böttcher C, Lindner P (2002) Structure of polyelectrolyte copolymer micelles. Macromolecules 35:4096
4.3 Förster S, Schmidt M, Antonietti M (1990) Static and dynamic light scattering by aqueous polyelectrolyte solutions: Effect of molecular weight, charge density and added salt. Polymer 31:781
4.4 Vrij A, Overbeek, JTG (1962) Scattering of light by charged colloidal particles in salt solutions. J Colloid Interface Sci 17:570
4.5 Tanford, C (1961) Physical chemistry of macromolecules. Wiley, New York

5 Selected Examples of Light Scattering Experiments

In this chapter, I will try to illustrate the practice of experimental light scattering by several examples chosen from recent literature. As a starting point, let us first consider a general flow chart on how a light scattering experiment typically is planned depending on sample information already available and sample characterization desired (see Fig. 5.1).

Note that the scheme shown in Fig. 5.1 is valid only in case all samples studied are very dilute, and interparticle interactions for this reason can be neglected. Only then is the pure particle form factor detected in a static light scattering experiment, and the relaxation process measured in a dynamic light scattering experiment corresponds to the selfdiffusion coefficient defined by the Stokes–Einstein–equation (Eq. 1.47). Particularly the effect of interparticle interactions on the experimental results of a DLS measurement will be discussed in more detail further below (see Sect. 5.1).

Next, let us consider the case of an unknown sample topology in more detail, as described by the flow chart presented in Fig. 5.2.

Here, it should be noted that whereas M_w, $\langle R_g \rangle_z$, A_2 and $\langle R_H^{-1} \rangle_z$ can be accurately measured by light scattering for most samples, determination of the sample topology itself often is a very difficult task. This is most obvious for smaller particles <50 nm, where measurement of the particle form factor is beyond the length-scale resolution of the light scattering experiment and therefore only the ρ -ratio R_g/R_H can provide an indication, but does not allow an unambiguous determination, of the particle shape. Also, in case of particles >50 nm, a large sample polydispersity may interfere with a detailed analysis of the particle form factor in a static light scattering experiment. Fractionation of the sample by GPC or FFF, and subsequent analysis of the less polydisperse sample fractions by light scattering in this case is the only way to achieve a reliable sample characterization, as already stated above.

The flow charts shown in Figs. 5.1 and 5.2 can only crudely summarize the experimental practice of light scattering. To illustrate the numerous applications of light scattering and also to provide detailed information on experimental procedure and data analysis of modern light scattering experiments, the author has tried to select a comprehensive set of representative examples from recent literature (1995 until now). All papers were written by widely accepted experts in

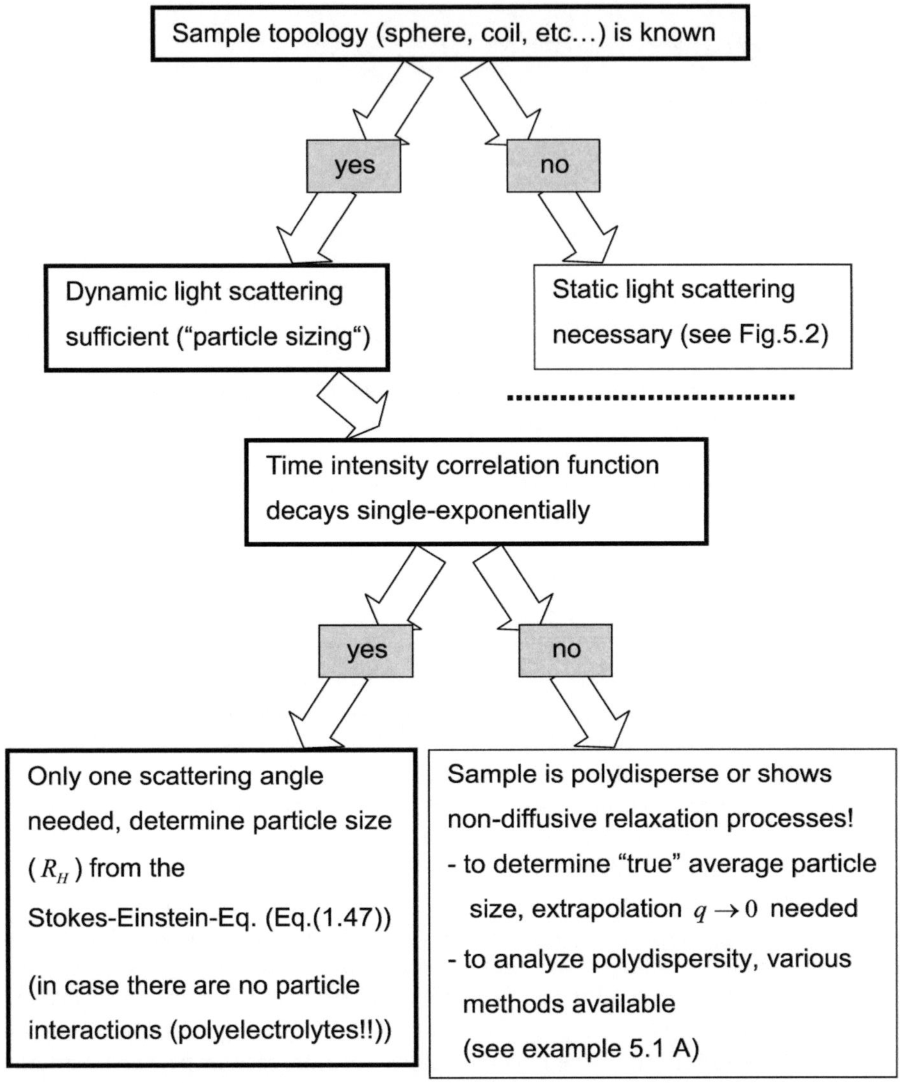

Fig. 5.1. Flow chart for sample characterization by laser light scattering – first part

the field of light scattering, and major contributors to the progress in application and experimental procedure of the light scattering method. We will now move on to a very important part of this book: a detailed review of examples from literature, presented in the context of the preceding chapters. All examples will be reviewed under the following aspects: (i) sample investigated and sample treatment before light scattering, (ii) light scattering setup used for the experiment, (iii) light scattering data and data analysis, (iv) concluding remarks.

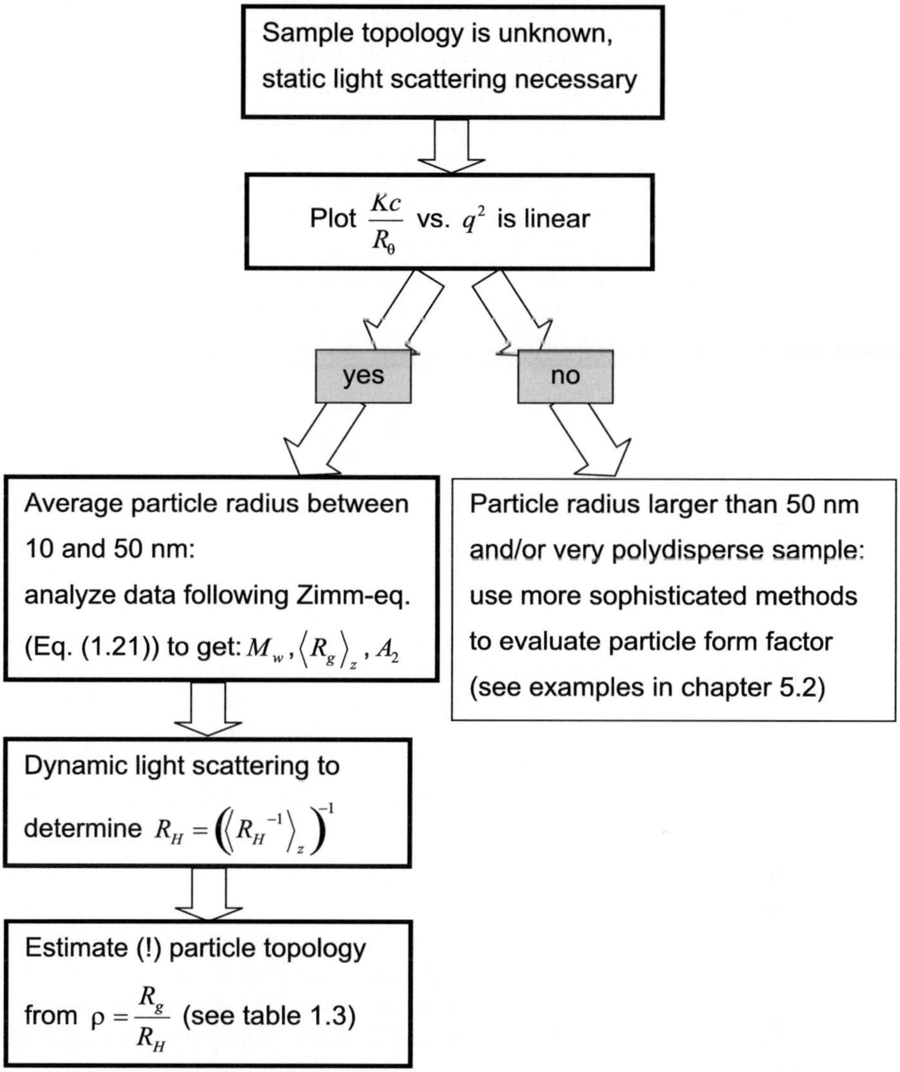

Fig. 5.2. Flow chart for sample characterization by laser light scattering – second part

In the first section of this chapter, several illustrative examples of the practice of pure dynamic light scattering studies are presented. One important aspect is the practice of polydispersity analysis. Another topic is the investigation of nonisotropic particles by dynamic light scattering. Also, the difficulties which may arise if solutions of charged scattering particles (= polyelectrolyte solutions) are investigated by dynamic light scattering are discussed.

In the second section, we will present several examples for light scattering studies of nanoscopic particles and polymer architectures of unknown topology. Here, often a combination of static and dynamic light scattering is used to characterize the unknown particles, as shown in the flow charts (Figs. 5.1, 5.2). Additionally, a couple of different approaches concerning the data analysis of the static scattering intensity or particle form factor are presented, depending on average particle size in respect to the experimental length scale and particle topology.

Finally, in Sect. 5.3 I will present some experimental examples for the new experimental dynamic light scattering techniques described in Chap. 2 of this book: fiber optic quasi elastic light scattering (FOQELS), dual color dynamic light scattering, and dynamic light scattering using a CCD camera chip as area detector. Here, the application of DLS to the characterization of turbid samples, as well as an enhancement of the time-scale of the dynamic light scattering method will be illustrated.

5.1 Dynamic Light Scattering

In this section, I will first provide an experimental example for the practice of polydispersity data analysis in dynamic light scattering experiments, using a spherical latex dispersion as model sample. This first example will be followed by examples for DLS from branched polymer architectures and rod-like particles. DLS experiments from charged particles will be presented thereafter. Section 5.1 will conclude with examples from the literature that show the treatment of usually unwanted side effects encountered during a dynamic light scattering experiment, as aggregate formation, presence of dust, and absorption of the incident laser light.

Example A (Ref. [2.1]):
The authors of ref. [2.1], Vanhoudt and Clauwaert, were mainly interested in investigating the difference between fiber optic and pinhole detectors used in light scattering experiments. Importantly, their study also provides a very nice example for state-of-the art data analysis of DLS data from polydisperse samples, covering all methods currently in use. For their experiments, Vanhoudt and Clauwaert have chosen a well-defined binary mixture of commercially available standard latex particles. In briefly reviewing their results, we will restrict ourselves to the classical pinhole detector setup. This restriction is justified by the conclusions of the authors who claim that, besides a lower weight and therefore easier experimental handling, fiber optic detectors do not improve the size resolution of a dynamic light scattering experiment compared to conventional pinholes.

(i) Samples and sample treatment:
Aqueous dispersions of commercial latex beads 3020A and 3050A Nanosphere Size Standards from Duke Scientific Corp. and one standard from Balzers Co.

were used in these experiments. Mean diameters of these particles as provided by the company and determined by DLS were 19 nm ± 1.5 nm, 54 nm ± 2.7 nm, and 91 nm ± 3 nm. Three different bimodal samples have been prepared from these nearly monodisperse standards such that the scattered intensities of the smaller and larger particle fraction detected at scattering angle 90° were identical within experimental error. All samples reviewed here are listed in Table 5.1.

We should note that a bimodal particle size distribution can be resolved in a DLS experiment only if the particle diameters of the two size fractions differ by at least a factor of 2. Correspondingly, the bimodal particle mixtures s6 and s7 have been chosen by the authors to test the limits of both experimental detector setups (which is not an issue here) and data treatment.

The cylindrical sample cells of 10 mm outer diameter, stored in an ethanol-HCl mixture if not in use, were rinsed with distilled water and further cleaned by flushing with condensing acetone vapour to remove dust. The samples were diluted with distilled water (filtered 3 times with 0.22-μm Millipore disposable filters) to adjust concentrations suitable for the DLS measurement. These concentrations were identified according to the mean scattered intensity detected at a scattering angle of 90°: 600 kHz for all monomodal samples and 120 kHz for all bimodal samples. These mean intensities or "count rates" are in the typical regime for DLS measurements (50–1,000 kHz). Here, it should be mentioned that for a given sample an older light scattering setup based on an Ar^+ ion laser and a photomultiplier will detect the same count rate in scattered intensity as a modern light scattering setup consisting of a highly stable 22 mW HeNe laser and a very sensitive avalanche photodiode (APD).

Note that at higher intensities the optical detector may be damaged after exposure of times longer than several seconds, whereas at lower intensities the signal-to-noise-ratio becomes too large for an accurate determination of the autocorrelation function. One should keep in mind that the scattered intensity in practice may strongly increase, if the detector is moved from larger to small (<50°) scattering angles. To avoid damage of the detector system, in angular-dependent light scattering studies always the average count rates at scattering angles covering the whole scattering angle range, typically between 150° and 30°, should be briefly and carefully monitored before the measurement. Here, carefully means the operator should be prepared to immediately block the scattered light, for example with a suitable shutter, if count rates larger 1,000 kHz are

Table 5.1. Monomodal and bimodal spherical latex particle dispersions studied in ref. [2.1], all sizes in nm. Reprinted with permission from J. Vanhoudt and J. Clauwaert, Langmuir 15, 44–57, 1999, Copyright 1999 American Chemical Society.

Sample	s2	ss3	s4	s5	s6	s7
Nominal diameter	19	54	91	19, 91	19, 54	54, 91
Diameter ratio	–	–	–	4.8	2.8	1.7

detected by the experimental system. For this purpose, all setups should, as commercial systems usually do, provide a monitor option to display the count rate currently detected.

To further reduce the amount of dust particles and disturbing particle aggregates, all samples were centrifuged for 30 min at 4,000 g using a Beckman Instruments Inc. machine right before the measurements.

(ii) Light scattering setup:
Vanhoudt and Clauwaert used a classical DLS setup, equipped with a vertically polarized Ar^{+}-laser (Spectra-Physics model 2017 Stabilite) operating at wavelength 488 nm as a light source. The laser beam was focused onto the sample by a biconvex lens, forming a scattering volume of about 100 μm in diameter. The sample itself was embedded in a glass container filled with filtered (0.45-μm Millipore disposable cartridge filters) water, serving as a temperature control and refractive index matching bath. The detector was positioned on the arm of an ALV goniometer and consisted of the standard ALV pinhole setup, which is part of the elder commercial ALV goniometers (ALV SP86), an amplifier-discriminator, and a Brookhaven Instruments BI-8000AT digital hardware correlator. This correlator covers up to 128 time channels and can be programmed to allow logarithmic sample time spacing with a total correlation time range between 1 μs and 150 ms. All measurements were performed at only one scattering angle (90°).

The reader should note that, as we have described in detail in Chap. 1.3, only apparent diffusion coefficients (and therefore apparent average particle radii) are detected in this single-angle DLS experiment. Only at $q = 0$ or scattering angle $\theta = 0$ do two scattering particles of the two different particle size fractions have a scattering power depending only on their particle mass, and as a consequence the correlation function measured by the DLS experiment reflects the z-average sample composition, the average diffusion coefficient in this case corresponding to a z-average. Since here we are interested only in comparing various methods of data treatment for a given data set, this aspect will be neglected but nevertheless should be kept in mind.

(iii) Methods used for data analysis:
The authors used the Brookhaven ISDA software packet to analyze the experimental correlation functions. This packet contains five different standard routines to determine the size distribution from time-intensity correlation functions measured for polydisperse samples.

In the following, I will in most cases use the same nomenclature as in ref. [2.1], but refer to the equations given in Chap. 1 of this book if necessary. The five alternative routines commonly used for analysis of DLS data from polydisperse samples are:

1. The Cumulant method (CUM), where the experimental data points are fitted to a polynomial series expansion according to:

$$\ln\left(f^{0.5}\left|g_1(\tau)\right|\right)=\ln\left(f^{0.5}\right)-\langle\Gamma\rangle\tau+\frac{\kappa_2\tau^2}{2} \tag{5.1}$$

where

$$\langle\Gamma\rangle=\int_0^\infty B(\Gamma)\Gamma d\Gamma \tag{5.2}$$

$$\kappa_2=\int_0^\infty(\Gamma-\langle\Gamma\rangle)^2 B(\Gamma)d\Gamma \tag{5.3}$$

Here, f is the contrast factor, and $\langle\Gamma\rangle$, κ_2 are the first- and second-order moments of the decay rate distribution $B(\Gamma)$. The polydispersity index, which is defined as $PI=\kappa_2/\langle\Gamma\rangle^2$, according to the Stokes–Einstein–equation must be identical for diffusion coefficient and inverse particle size distribution. The average decay rate $\langle\Gamma\rangle$ is a z-average in the limit $q=0$. At finite scattering angles, it leads to the apparent selfdiffusion coefficient according to $\langle\Gamma\rangle=D_{app}\cdot q^2$. The associated particle diameter, calculated using the Stokes–Einstein–equation, according to the authors of ref. [2.1] is a so-called harmonic z-average defined as

$$\langle d\rangle=\frac{\sum_i n_i d_i^6}{\sum_i n_i d_i^5} \tag{5.4}$$

with n_i the number of particles with diameter d_i. The reader should note that Eq. 5.4 is derived for spherical particles only, in which case the particle mass scales with particle size as $M_i\sim d_i^6$. The harmonic average lies in between the z-average and the weight-average particle size. On the other hand, if we consider the average inverse particle size $\langle R_H^{-1}\rangle$, this value if extrapolated to $q=0$ is a z-average (see Chap. 1.3). Here, it should be kept in mind that the apparent selfdiffusion coefficient measured at a finite scattering angle usually leads to an apparent average particle size smaller than corresponding to the "true" inverse z-average. Finally, it should be noted that for bimodal samples consisting of two well-defined monodisperse size fractions as considered here, the mean particle size and the polydispersity index defined in the Cumulant method have no physical meaning but are rather arbitrary parameters to characterize the sample polydispersity.

2. The nonnegatively least squares method (NNLS), which calculates a histogram of the particle size distribution by minimizing the following expression:

$$\chi^2=\sum_{j=1}^{N}\left[g_1(\tau_j)-\sum_{i=1}^{M}b_i\exp\left(-\Gamma_i\tau_j\right)\right]^2 \tag{5.5}$$

with N the number of correlation times or channels constituting the experimental correlation function $g_1(\tau)$, and M the number of exponentials considered for the series expansion. This approach yields a set of coefficients b_i defining the particle size distribution for a given set of decay rates Γ_i, which are equally distributed on a logarithmic scale in a range chosen by the operator.

3. Exponential sampling (ES) is an alternative method to calculate a discontinuous decay rate distribution function given by the set $\{b_i, \Gamma_i\}$ (see Eq. 5.5). Here, the set of decay rates Γ_i is defined according to:

$$\Gamma_{n+1} = \Gamma_1 \exp\left(n\pi / \omega_{max} \right) \tag{5.6}$$

ω_{max} is a fit parameter determined by trial and error, and the coefficients b_i are calculated by minimization of the expression given above (Eq. 5.5).

4. Based on Provencher's CONTIN algorithm [5.1], a solution of the Laplace transform of the correlation function $g_1(\tau)$ and therefore the decay rate distribution $B(\Gamma)$ is obtained by minimization of the following expression:

$$\chi^2(\alpha) = \sum_i \left(1/\sigma_i^2 \right) \left[g_1(\tau_i) - \int B(\Gamma) \cdot e^{-\Gamma\tau} d\Gamma \right]^2 + \alpha \left\| \hat{L} B(\Gamma) \right\|^2 \tag{5.7}$$

σ_i is the standard deviation for the experimental data point $g_1(\tau_i)$, and $\alpha \left\| \hat{L} B(\Gamma) \right\|^2$ the so-called regulation term, with $0 < \alpha < 1$ defining the detail of the solution of the equation, and the operator $\hat{L}$ typically chosen as second derivative. This method is not recommended for multimodal size distributions as the one treated here, but gives good results for monomodal polydisperse samples, in which case it also allows a good estimate of the size polydispersity.

5. The double-exponential method (DE) treats the sample as a composition of two well-defined monodisperse species, and therefore assumes a model autocorrelation function given by the sum of two exponentials:

$$g_1(\tau) = b_1 e^{-\Gamma_1 \tau} + b_2 e^{-\Gamma_2 \tau} \tag{5.8}$$

The parameters $b_{1,2}$ and $\Gamma_{1,2}$, respectively, are determined by a least-squares fit of this model function to the experimental data, yielding contributions to the scattered intensity defining the size distribution ($b_{1,2}$), and size of the two sample species ($\Gamma_{1,2} = D_{s1,2} \cdot q^2$). Obviously, this method is well suited to characterize bimodal particle size distributions. However, it should be stressed that even in this simple case the dependence of the correlation function on scattering angle cannot be ignored, since the ratio b_1/b_2, that is the relative scattered intensities, may vary with q due to the effect of the particle form factor.

(iv) Experimental results:
Results from the different analysis methods are summarized in Table 5.2.

Table 5.2. Sample (see Table 5.1) characteristics for various DLS data treatment methods, all particle sizes *d* are given in nm. PI is the polydispersity index. Reprinted with permission from J. Vanhoudt and J. Clauwaert, Langmuir 15, 44–57, 1999, Copyright 1999 American Chemical Society.

Sample	s2	s3	s4	s5	s6	s7
Nominal diameter	19 ± 1.5	54 ± 2.7	91 ± 3	19, 91	19, 54	54, 91
Diameter ratio	–	–	–	4.8	2.8	1.7
$\langle d \rangle$ – CUM	20.3	55.0	87.0	36.9	29.5	69.0
PI – CUM	0.029	0.009	0.008	0.248	0.191	0.069
d_1, d_2 – NNLS	–	–	–	18.2, 80.7	16.4, 50.1	–
b_1, b_2 – NNLS	–	–	–	0.45, 0.55	0.42, 0.58	–
d_1, d_2 – ES	–	–	–	–	19.0, 54.0	–
b_1, b_2 – ES	–	–	–	–	0.45, 0.55	–
d_1, d_2 – DE	–	–	–	–	18.1, 53.7	–
b_1, b_2 – DE	–	–	–	–	0.45, 0.55	–

Regarding first the monomodal samples listed in Table 5.2 (s2, s3, s4), we find that in Cumulant analysis PI-values smaller than 0.01 correspond to samples of size polydispersity $\Delta d/\langle d \rangle$ less than 5% (samples s3, s4). Consequently, for the bimodal samples (s5, s6, s7) the Cumulant analysis yields PI-values larger than 0.05 in all cases. Interestingly, the average particle diameters determined by Cumulant analysis for the bimodal samples s5 and s6 seem to be too small compared to the theoretical sample composition. For sample s7, the average particle diameter seems to be in agreement with the expected value. Here, one has to take into account, as mentioned before, that at finite scattering angles >0° the contribution of the smaller particle size fraction to the scattered intensity is comparatively larger than that of the larger size fraction, leading in a DLS experiment to an apparent average particle size smaller than the "true" inverse z-average.

Concerning the accurate measurement of the size distribution of bimodal samples by DLS, sample s6 is very difficult to analyze, since here the diameter ratio of the two particle species is close to the critical value of 2, the resolution limit for DLS. In this case, the NNLS routine yields the wrong two average particle sizes, whereas ES and DE give similar results within 5% deviation from the true values and in even better agreement with the size of the larger particle fraction.

A general problem of the characterization of bimodal samples by DLS is the accurate determination of the size of smaller particles in the presence of larger ones, whose scattering intensity is much stronger since $\langle I \rangle \sim d^6$: for sample s5, where the resolution of the bimodality should not cause a problem due to the large particle size ratio of 4.8, NNLS already leads to comparatively bad results.

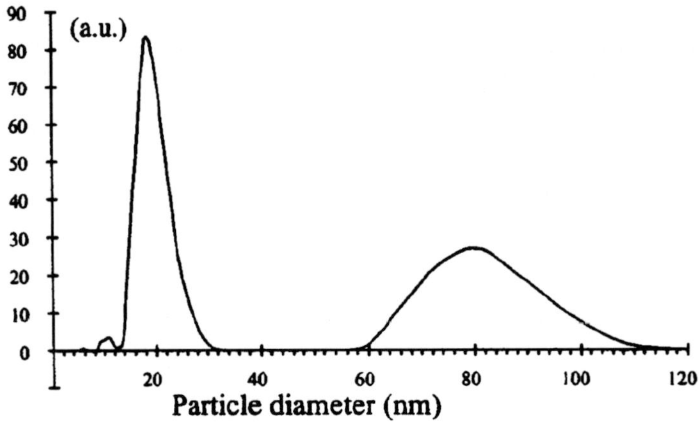

Fig. 5.3. Size distribution of bimodal sample 5 determined by the NNLS analysis. Reprinted with permission from J. Vanhoudt and J. Clauwaert, Langmuir 15, 44–57, 1999, Copyright 1999 American Chemical Society (ref. [2.1]).

Figure 5.3 shows the size distribution function determined by the NNLS-method for sample s5.

The corresponding two average particle sizes and peak intensities are found in Table 5.2. On the other hand, the bimodality of sample s7 with a size ratio well below 2 is beyond the resolution of dynamic light scattering.

(v) Conclusions:

The example shows the meaning of the Cumulant data analysis for the practice of dynamic light scattering characterization of polydisperse samples. It also illustrates that bimodal size distributions can be resolved if the size ratio is larger than 2, but size distribution analysis fails for bimodal samples with smaller size ratios. Finally, we have seen that exponential sampling (ES), using an adjustable exponential range of relaxation times, in case of a bimodal sample leads to more reliable results than the standard NNLS routine where relaxation times are equally distributed on a log-scale.

Example B (Ref. [5.2]):

Our next example has been chosen to illustrate dynamic light scattering from nonspherical or noncompact objects, in which case internal modes of particle mobility, like polymer segment motion, contribute to the correlation function, and therefore have to be taken into account during data analysis if one is just interested in particle sizing. Further, these internal modes themselves may provide interesting information concerning the sample characteristics. A possibility to separate the contributions from internal modes and particle selfdiffusion to the decay rate of the correlation function is provided by the length scale of the light scattering experiment, as will be shown in this example.

(i) Samples and sample treatment:
Galinsky and Burchard studied 0.5 M NaOH aqueous solutions of degraded starch as an example of randomly branched polymeric nanoparticles. The molecular parameters of the samples determined by both static and dynamic light scattering from highly dilute solutions are listed in Table 5.3.

All solutions were filtered three times through Millipore filters of pore size 1.2, 0.8, 0.45, or 0.2 μm depending on average particle size. Typically, one should choose a pore size at least twice the average size of the scattering particles to avoid clogging of the filter. The third time the samples were filtered directly into 0.8-cm inner diameter cylindrical light scattering cells. Before use, these cells were rinsed with freshly distilled acetone in a special apparatus to remove dust particles.

(ii) Light scattering setup:
The setup was an ALV instrument equipped with the linear ALV3000 correlator (linear scale of correlation times) and the nonlinear ALV5000 correlator (pseudo-logarithmic scale of correlation times, obtained by doubling the timescale after every 8 channels). These two correlators only yield identical apparent diffusion coefficients if certain conditions are met, as will be shown in part (iii).

(iii) Methods used for data analysis:
The authors determined the apparent selfdiffusion coefficient via a series expansion according to:

$$D_{app}(q) = \frac{\Gamma}{q^2} = D_z\left(1 + C_H\left(q^2 R_g^{\ 2}\right)\right) \tag{5.9}$$

D_z is the z-average diffusion coefficient obtained by extrapolation of the apparent diffusion coefficient towards zero scattering angle ($\theta = 0$), and C_H a coefficient sensitive to the particle topology, which will be discussed in more detail in Sect. (iv).

To analyze the angular dependence of the decay rates $\Gamma(q)$ in detail, a combination of the DLS data obtained with the linear and the nonlinear ALV correlators had to be used. Here, different results for $D_{app}(q)$ are obtained, if certain

Table 5.3. Sample characteristics of hyperbranched degraded starch molecules. Reprinted with permission from G. Galinsky and W. Burchard, Macromolecules 30, 6966–6973, 1997, Copyright 1997 American Chemical Society (ref. [5.2]).

Sample	$10^{-6} M_w$ [g/mol]	R_g [nm]	$10^5 A_2$ [mol mL/g^2]	$10^8 D_z$ [cm^2/s]	R_H [nm]	ρ
LD16	1.7	48	6.00	6.2	36	1.37
LD12	5.2	70	2.00	3.5	62	1.14
LD19	14.5	113	1.30	2.6	84	1.35
LD18	43.0	180	0.82	1.4	150	1.20
LD17	64.0	190	0.60	1.0	215	0.88

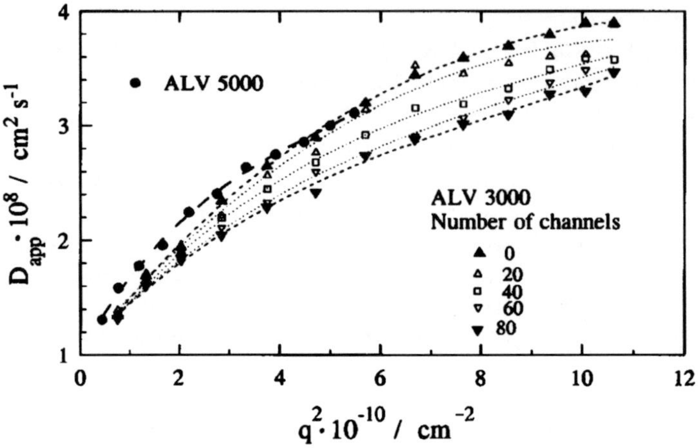

Fig. 5.4. Effect of the hardware correlator type and the number of channels of the analyzed correlation functions on the q^2-dependence of apparent diffusion coefficients for branched macromolecules (starch). Reprinted with permission from G. Galinsky and W. Burchard, Macromolecules 30, 6966–6973, 1997, Copyright 1997 American Chemical Society (ref. [5.2]).

conditions are not met. Galinsky and Burchard analyzed their DLS data based on the Cumulant series expansion of the amplitude correlation function, given as:

$$\ln g_1(\tau,q) = \Gamma_0 - \Gamma_1\tau + \left(\frac{\Gamma_2}{2!}\right)\tau^2 - \left(\frac{\Gamma_3}{3!}\right)\tau^3 + \ldots \tag{5.10}$$

Conditions for consistent solutions were: first, all Cumulants had to be positive. Second, $\Gamma_2/2! \approx \Gamma_3/3! \cdot \tau$. These conditions were met if, in case of the nonlinear ALV5000 correlator, the correlation functions had not decayed by more than 30% of the initial value, and, in case of the linear ALV3000 correlator, the number of channels had been reduced to 20. Here, we have an example that the channel number, which is the correlation time window used for the Cumulant analysis, may have a strong influence on the experimental results obtained from the Cumulant analysis, as is illustrated by the data shown in Fig. 5.4.

The apparent diffusion coefficient at given scattering vector q increases systematically with decreasing channel number for channel numbers larger than 20. This behavior is easy to understand if one considers that the Cumulant approach is a series expansion (see Eq. 5.10) and therefore only valid for comparatively small correlation times τ. In the present example, Γ_1 was considered as describing the "true" apparent diffusion coefficient in case the number of channels of the ALV3000 correlation functions was smaller than 20.

(iv) Experimental results:

Figure 5.5 shows that the authors could superimpose all their data for $D_{app}(q)$ of the different starch samples by normalization with D_z and R_g^2, respectively, to

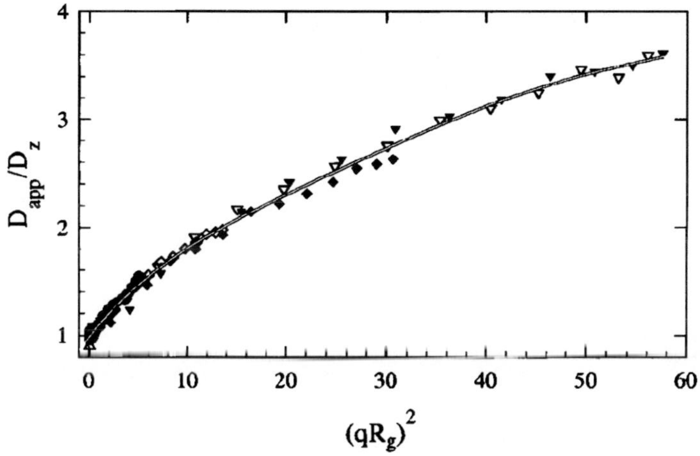

Fig. 5.5. Scaled master-function for the q^2-dependence of apparent diffusion coefficients for branched macromolecules (starch). Reprinted with permission from G. Galinsky and W. Burchard, Macromolecules 30, 6966–6973, 1997, Copyright 1997 American Chemical Society (ref. [5.2]).

obtain a universal master curve for all five sample fractions irrespective of their different molar masses.

The initial slope of the plot shown in Fig. 5.5 according to Eq. 5.9 corresponds to the coefficient C_H and lies between 0.102 and 0.134. The authors consider the existence of this universal behavior as highly unexpected, since the particle form factors $P(q)$ of their samples measured in a static light scattering experiment could not be superimposed.

Importantly, the angular dependence of the apparent diffusion coefficients shown in Figs. 5.4 and 5.5 is not caused by polydispersity (see Chap. 1.3 of this book), but by internal segmental motions which lead to a q-independent contribution to the decay rate of the correlation function $g_1(q,\tau)$. A more detailed discussion of the relaxation processes is based on the DLS data representation shown in Fig. 5.6, a double logarithmic plot of the normalized apparent diffusion coefficient vs. qR_g.

At $qR_g > 3$, that is at small length scales compared to the size of a single scattering particle, no translational diffusion but only internal relaxation processes such as rotation, segment motion etc., should determine the relaxation processes of the correlation function, and therefore contribute to $D_{app}(q)$. Instead of the q^1-dependence (slope 1 of the data plotted in Fig. 5.6) expected for the relaxation of Gaussian chain segments and found for the linear chain reference system, the branched starch samples show a slope below 0.85, corresponding to a weaker q-dependence. In addition, the data for the starch samples which are nonrandomly branched are shifted to higher qR_g compared to the results for a randomly branched polyester system with much lower branching density. Importantly, all data from the starch samples fall on a single master curve, indicating their universal dynamic behavior. In Fig. 5.6, the shift of the data to comparatively larger qR_g-values

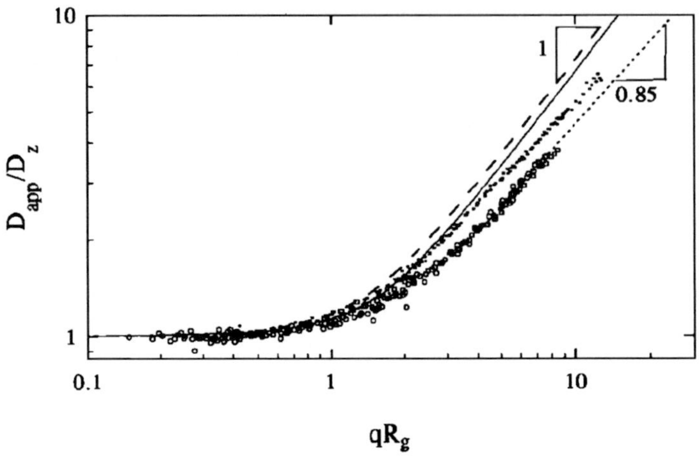

Fig. 5.6. Double-logarithmic plot of normalized D_{app} vs. qR_g for starch samples, including also plots for branched polyesters and linear chains for direct comparison. (Reprinted with permission from G. Galinsky and W. Burchard, Macromolecules 30, 6966–6973, 1997, Copyright 1997 American Chemical Society (ref. [5.2]).

shows that internal segmental motions in branched systems are detected on a shorter length scale (= larger q) than in Gaussian linear polymer chains or randomly branched polymer samples.

A more detailed analysis of the dynamic properties of the branched polymers is obtained if we regard the so-called normalized first Cumulant defined as:

$$\Gamma^*(q) \equiv \left(\frac{\Gamma}{q^3}\right)\left(\frac{\eta_0}{kT}\right) = \left(\frac{D_{app}(q)}{q}\right)\left(\frac{\eta_0}{kT}\right) = \frac{1}{6\pi q \xi_h(q)} \tag{5.11}$$

Here, $\xi_h(q)$ is the so-called hydrodynamic correlation length not further specified by Galinsky and Burchard. Figure 5.7 shows experimental results for the normalized diffusion coefficient or first Cumulant $\Gamma^*(q)$ for the starch samples, together with the normalized first Cumulants of a linear polymer coil and of hard sphere colloids as two references.

The data for the hyperbranched starch samples fall in between the two limiting cases of linear chains and hard spheres. This is plausible, since the starch samples can be regarded as intermediate structures between the two extreme reference systems. In Fig. 5.7, a decrease in decay rate $\Gamma^*(q)$ (compared to the linear chain reference) at fixed q, that is at fixed length scale, corresponds to an increase in relaxation time and therefore a slowing down of the mobility of polymer chain segments of the same length. The data are interpreted such that branching in a polymer system leads to a loss in chain flexibility. Moreover, the authors also conclude from the shape of $\Gamma^*(q)$ that the relaxation spectrum of internal motions itself is changed for branched systems compared to linear chains. However, this discussion, which is based on advanced polymer physics, seems to be beyond the scope of this book focused on the light scattering technique.

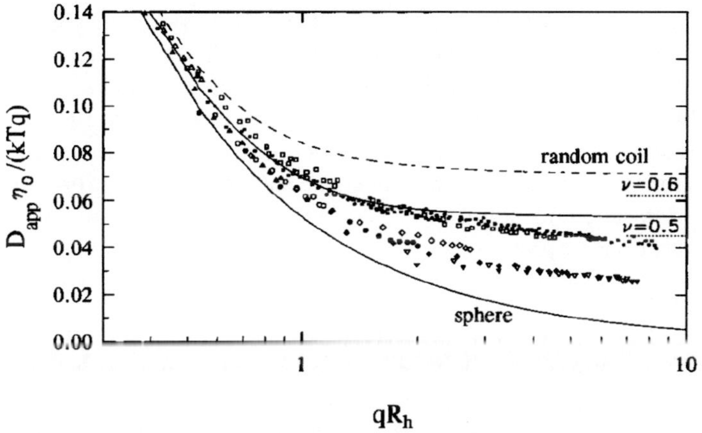

Fig. 5.7. q-dependence of normalized first Cumulant $(D_{app}(q)/q)(\eta_0/kT)=\Gamma^*(q)$ for starch samples, and for linear polymers and hard spheres. Reprinted with permission from G. Galinsky and W. Burchard, Macromolecules 30, 6966–6973, 1997, Copyright 1997 American Chemical Society (ref. [5.2])

(v) Conclusions:
This paper provides a very nice example of how internal modes like polymer chain segment motion influence the experimental data in a dynamic light scattering experiment. It also illustrates that dynamic light scattering is not just a tool for particle sizing by measuring the Stokes–Einstein diffusion coefficient, but also can yield detailed information on the complex internal dynamics of systems deviating from the simple case of solid spheres, like flexible linear or branched polymer samples, or rod-like colloidal particles. The later will be reviewed in examples C to E of this section.

Example C (Ref. [5.3]):
Solutions of charged samples provide a major challenge for all light scattering experiments. If one is interested in particle characterization of such systems, it is mandatory to suppress any particle interactions caused by Coulomb forces and strongly influencing the experimental data both in static and dynamic light scattering experiments. Our next example will try to describe the effect of charge interactions on dynamic light scattering results for the interesting case of an aqueous solution of DNA, which is also an example for dynamic light scattering from anisotropic particles (for example flexible or stiff rods).

(i) Samples and sample treatment:
Commercially available high molar mass calf thymus DNA (Sigma D-1501) of approximate length of 400 base pairs was used for the studies by Borsali, Nguyen, and Pecora. The DNA was dissolved in a buffer solution and ultrasonicated at 10 kHz and at a temperature of 0°C for 5 min. Residual protein was removed from

the DNA buffer solution by repeated phenol-chloroform extraction, and phenol still present in the solution was removed by ether extraction. The ether itself was evaporated in vacuum. The thusly purified DNA was precipitated in ethanol and redispersed in double-distilled water at a concentration of 50 mg/mL as verified by uv absorption at 260 nm. Any residual salt (NaCl) was removed by repeated dialysis versus double distilled water over 4 days to ensure a salt-free system. From this stock salt-free solution, aqueous solutions of defined salt concentration in the regime 5 – 1000 mM NaCl were prepared.

(ii) Light scattering setup:
The apparatus consisted of a Spectra Physics Model 165 argon ion laser operating at wavelength 488 nm as the light source. Scattered light was detected at scattering angles 30° to 130°, corresponding to a scattering vector regime 0.65e-3 $Å^{-1} < q <$ 3.01e-3 $Å^{-1}$. The laser power was between 100 and 500 mW, and the scattered intensity fluctuations were correlated with a Brookhaven BI9000 hardware autocorrelator.

(iii) Methods used for data analysis:
The distribution of decay rates $A(\Gamma)$ was determined from the correlation function $g_1(q,\tau)$ using the CONTIN algorithm (see example A) according to:

$$\left[\frac{g_2(q,\tau)}{B}-1\right]^{0.5} = g_1(q,\tau) = \int_0^\infty A(\Gamma)e^{-\Gamma\tau}d\tau \tag{5.12}$$

A typical example decay rate distribution $A(\Gamma)$ for the DNA solution containing the highest salt concentration investigated in ref. [5.3] is shown in Fig. 5.8.

Alternatively and, according to the authors, more reliably, the correlation functions have also been analyzed using the double-exponential fit method best suited for samples with a bimodal size distribution (see example A).

(iv) Experimental results:
The decay rates shown in Fig. 5.8 exhibit a bimodal decay rate distribution corresponding to a fast and a slow process. It should be noted that at high salt concentrations (1 M NaCl) Coulomb interactions between the scattering particles should be negligible due to screening. To understand the origin of the two processes in detail, the authors explored the dependence of decay rates on scattering angle or experimental length scale in some detail, as shown in Fig. 5.9.

Both relaxation processes determined from a CONTIN analysis of the correlation function $g_1(q,\tau)$ show a linear dependence of decay rates on q^2 as expected for translational diffusion, and the slope of these lines corresponds directly to the selfdiffusion coefficients according to $\Gamma = Dq^2$. The DNA concentration itself seems to have a negligible influence on the particle dynamics at concentrations up to 0.66 mg/mL as also shown in Fig. 5.9.

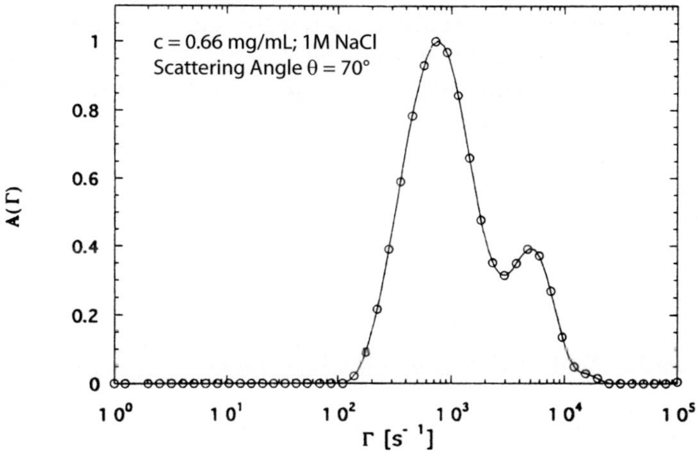

Fig. 5.8. Decay rate distribution for a high-salt DNA sample. Reprinted with permission from R. Borsali, H. Nguyen and R. Pecora, Macromolecules 31, 1548–1555, 1998, Copyright 1998 American Chemical Society (ref. [5.3]).

The two diffusive relaxation processes correspond to diffusion coefficients $D_{fast} = 1.34 \cdot 10^{-7}\, cm^2 s^{-1}$ and $D_{slow} = 2.57 \cdot 10^{-8}\, cm^2 s^{-1}$, respectively. The reader should note that the two processes differ by a factor of 5, which is much larger than the minimum factor of 2 necessary to resolve a bimodal decay rate distribution in a dynamic light scattering experiment. Borsali et al. attributed the fast process to concentration fluctuations based on selfdiffusion of single DNA molecules. The

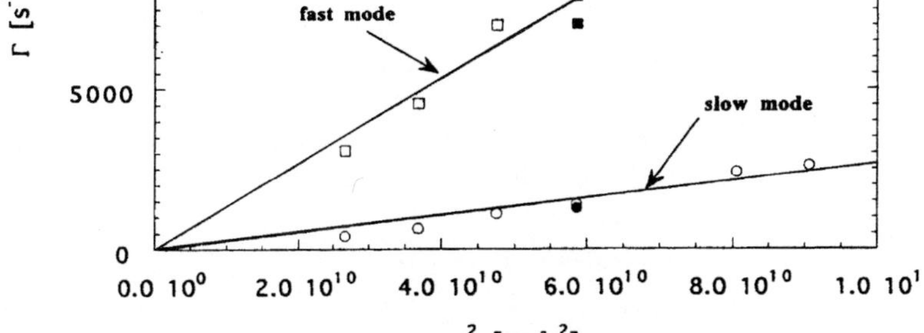

Fig. 5.9. q^2-dependence of relaxation processes for an aqueous DNA-solution. Reprinted with permission from R. Borsali, H. Nguyen and R. Pecora, Macromolecules 31, 1548–1555, 1998, Copyright 1998 American Chemical Society (ref. [5.3]).

corresponding hydrodynamic radius calculated from the Stokes–Einstein–equation is 159.6 Å. If the DNA molecule is assumed to have the topology of a semiflexible chain, its contour length L can be calculated from the dynamic light scattering result in case the persistence length P, which is a measure for the chain stiffness, is known. If a rod is perfectly rigid, that is $L \ll P$, according to Broersma [5.4] the selfdiffusion coefficient is given as:

$$D = \frac{kT}{3\pi\eta_0 L}\left[\ln(2p) + \frac{1}{2}(\gamma_\parallel + \gamma_\perp)\right] \tag{5.13}$$

with p the ratio of rod length L to cross sectional diameter d ($d = 20.5$ Å for DNA), and the other parameters of the equation defined as:

$$\begin{aligned}\gamma_\parallel &= -0.193 + \frac{0.15}{\ln(2p)} + \frac{8.1}{\ln^2(2p)} - \frac{18}{\ln^3(2p)} + \frac{9}{\ln^4(2p)} \\ \gamma_\perp &= 0.807 + \frac{0.15}{\ln(2p)} + \frac{13.5}{\ln^2(2p)} - \frac{37}{\ln^3(2p)} + \frac{22}{\ln^4(2p)}\end{aligned} \tag{5.14}$$

The origin of these two equations is a separation of the diffusion of rod-like particles into two different modes: a faster diffusion process parallel to the rod axis and a slower diffusion process perpendicular to the rod axis. The resulting equations are series expansions of expressions obtained from complex hydrodynamic model calculations which are far beyond this book. An alternative expression for the diffusion coefficient of rigid rods has been derived by Tirado and Garcia de la Torre [5.5]:

$$D = \frac{kT}{3\pi\eta_0 L}(\ln p + \nu) \tag{5.15}$$

with the parameter

$$\nu = 0.312 + \frac{0.565}{p} - \frac{0.1}{p^2}. \tag{5.16}$$

Again, p is the ratio of rod length L to rod diameter d. With the experimentally determined diffusion coefficient $D_{fast} = 1.34 \cdot 10^{-7}\, cm^2 s^{-1}$, the above relations yielded for the DNA sample a contour length L = 1,396 Å (Broersma) and L = 1,432 Å (Tirado). Both results agree within experimental error. On the other hand, assuming a semiflexible rod with persistence length P = 500 Å, a value commonly accepted for DNA in water at high salt content, the authors calculated a contour length of L = 1,632 Å, using the hydrodynamic model developed for worm-like chains by Fuji and Yamakawa [5.6].

The origin of the slow process is much more difficult to identify and, according to the authors, has been the reason for much debate and controversy. The hydrodynamic radius, calculated from $D_{slow} = 2.57 \cdot 10^{-8}\, cm^2 s^{-1}$ using the Stokes–Einstein–equation, is 835 Å, which corresponds to the typical size of about 1,000 Å found for the slow diffusion coefficient detected in many aqueous polyelectrolyte solutions. Most of those measurements from other research groups, however, were conducted under salt-free conditions, where Coulomb interactions are not screened, as they should be in the case of c(NaCl) = 1 M considered here. The most common interpretation of the "slow mode" found in the recent literature on dynamic light scattering of polyelectrolytes in salt-free solution assumes the existence of either temporal or permanent aggregates (or clusters). Some authors also ascribe these slow relaxation processes to sample regions of particle density higher than average, also called "dense domains." However, the authors Borsali, Nguyen, and Pecora, although known as experts in the field of light scattering, did not dare to speculate further on the physical origin of a "slow mode" found in their special case of DLS measurements of charged DNA in aqueous solution at high salt concentration. One possible reason for the slow mode occurring here at such high concentration of added salt might be so-called salting out: if the Coulomb interactions stabilizing a dispersion of charged colloidal particles are highly screened, the particles may form aggregates or even precipitate.

Here, it should be mentioned that very recent dynamic light scattering studies of aqueous solutions of well-defined poly(styrenesulfonate) chains by Paul Russo and coworkers indicate that the slow mode is due to temporal aggregates which are part of a thermodynamically equilibrated state in case of low added salt concentration [5.7]. Importantly, the salt concentration during these dynamic light scattering studies was adjusted by dialysis, leading to the establishment of a Donnan equilibrium (see Chap. 4.1).

To investigate the effect of nonscreened Coulomb interactions on the respective relaxation processes, Borsali et al. have also studied a salt-free DNA solution with dynamic light scattering. Unfortunately, these measurements were conducted in the highly concentrated regime c(DNA) = 42 mg/mL, leading to increasing complexity of the system. As a consequence, the authors discovered three relaxation processes. The corresponding correlation functions for "salt-free" and "high-salt" conditions are shown in Fig. 5.10.

As can be seen, the fastest relaxation mode, with $D_1 = 1.33 \cdot 10^{-7}\, cm^2 s^{-1}$, nearly disappears at salt-free conditions, is independent of DNA concentration, and corresponds to the single molecule diffusion process already discussed above, allowing particle sizing. The second mode corresponds to $D_2 = 6.5 \cdot 10^{-10}\, cm^2 s^{-1}$ for salt-free conditions and $D_2 = 7.5 \cdot 10^{-10}\, cm^2 s^{-1}$ for 1 M added salt. Comparing these values to $D_{slow} = 2.57 \cdot 10^{-8}\, cm^2 s^{-1}$ given above for the dilute solution (see Fig. 5.9), we see a strong effect of DNA concentration on this process but a comparatively weak influence of added salt. Here, it should be mentioned that Borsali et al. believe their intermediate relaxation process 2 to be identical to the slow relaxation process found in very dilute solutions at 1 M salt concentration. Unfortunately, the authors failed to show experimental DLS data obtained for salt-free solutions at low DNA concentration comparably to the samples presented in Figs. 5.8 and 5.9. Finally, the concentrated sample according to Borsali et al. also shows a third ultraslow process with $D_3 = 3.56 \cdot 10^{-12}\, cm^2 s^{-1}$ for salt-free conditions and $D_3 = 1.71 \cdot 10^{-12}\, cm^2 s^{-1}$ for 1 M added salt, which again seems not to depend too much on salt concentration. The authors attribute this third process to even larger aggregates, which seems to be plausible considering the high

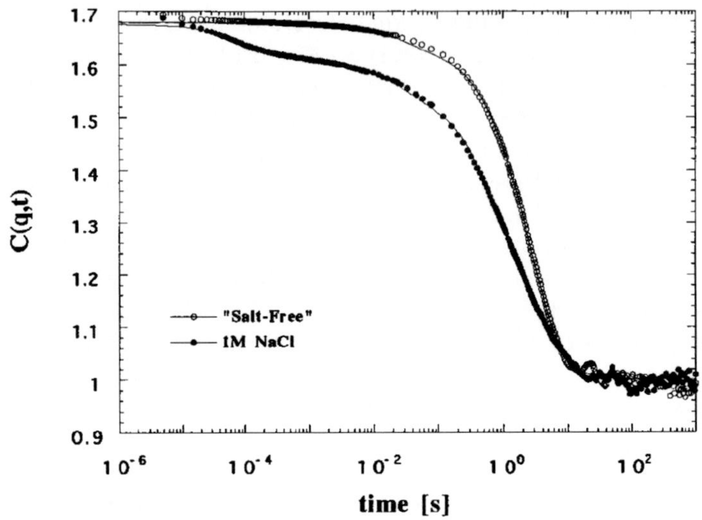

Fig. 5.10. Effect of added salt on correlation functions measured by DLS of DNA solutions in the semidilute regime (c(DNA) = 0.42 mg/mL) at scattering angle 90°. Reprinted with permission from R. Borsali, H. Nguyen and R. Pecora, Macromolecules 31, 1548–1555, 1998, Copyright 1998 American Chemical Society (ref. [5.3]).

concentration of the samples. However, one also should consider that the two slow processes D_2 and D_3 are separated by less than 2 orders of magnitude, which in the long-time regime of dynamic light scattering might be insufficient to clearly resolve the bimodality. More likely, the seemingly two processes D_2 and D_3 correspond to one single highly polydisperse relaxation process, since it is difficult to understand why two different and well-distinguishable species of small and large clusters should be formed in solution, leading to a bimodal decay rate distribution.

(v) Conclusions:
Example C shows that aqueous solutions of charged nanoparticles or polymers may give rise to DLS signals difficult to interpret, which sometimes is the case even in the presence of high amounts of salt added to screen the Coulomb interactions. Here, compared to the data presented in ref. [5.3], a more systematic study of the effects of both salt and DNA concentration on the relaxation processes detected in the DLS measurement would have been interesting, as already mentioned above. In respect to single particle characterization, the topic of our book, the reader should keep the following in mind: to measure single particle diffusion, as defined by the Stokes–Einstein–equation (Eq. 1.47), in a DLS experiment, a low particle concentration is crucial. Note that in case of charged particles (polyelectrolytes), low concentration may mean even lower than 0.1 g/L. Only in this case, screening of the Coulomb interparticle interactions by addition of salt allows one to distinguish the single particle relaxation process from the polyelectrolyte slow mode, a second relaxation process found in many DLS experiments on polyelectrolyte

samples and still not understood in general. Therefore, if one is not sure about the concentration regime suitable for single particle characterization while measuring DLS from polyelectrolyte solutions, it is absolutely necessary to systematically vary the sample concentration at given concentration of added salt (typically 0.001 Mol/L) until the DLS correlation functions remain unchanged, which for linear polyelectrolyte chains is the case at a ratio molar monomer concentration to molar salt concentration smaller than 1 (see Chap. 4). Here, special care is necessary in case of hydrophobic polyelectrolytes, which may show aggregation or even precipitation if too much salt is added to the particle solution ("salting out"). Finally, example C illustrates how to analyze DLS data obtained from anisotropic particles, a topic which will be further considered in the next two examples.

Example D (Ref. [5.8]):
Here, another example of dynamic light scattering from charged anisotropic nanoparticles in aqueous dispersion is presented: studies of the tobacco mosaic virus, which provides a well-defined monodisperse model system for stiff rods. One major difference compared to example C is not only the sample itself, but also the experimental method: Lehner, Lindner and Glatter have used depolarized dynamic light scattering to detect the rotational diffusion coefficient in addition to the translational particle motion. Last but not least, this example will give us some more insight in successful screening of unwanted charge effects by adding salt to the aqueous sample solution.

(i) Samples and sample treatment:
Tobacco mosaic virus (TMV) represents a well-suited model system for dynamic light scattering: it is a stiff cylindrical particle with length $L = 300$ nm, diameter $d = 18$ nm and molar mass $M = 4.0$ e7 g/Mol. Importantly, all particles are identical in size and shape, leading to a well-defined monodisperse light scattering sample, in case end-to-end aggregation of the TMV cylinders can be neglected. The TMV concentration of the aqueous sample solution was 0.25 g/L to exclude direct particle interactions. To check the influence of long-range Coulomb repulsion, the TMV particles were dissolved in pure doubly distilled water, and in 0.002 mol/L NaCl solution. No details how the samples have been cleaned from dust have been given in ref. [5.8].

(ii) Light scattering setup:
The measurements were performed with a laboratory-built goniometer equipped with an argon ion laser (Spectra Physics 2060-55), operating at wavelength 514.5 nm, as light source. Typical laser power for polarized light scattering (vv = "vertical vertical" geometry) was 150 mW; that for depolarized light scattering (vh = "vertical horizontal") was substantially higher (800 mW). Here, v and h indicate the polarization direction of the linearly polarized primary and scattered laser light with respect to the scattering plane. For a depolarized light scattering experiment, the incident laser beam typically has a vertical polarization whereas the scattered light has to be detected in horizontal polarization. Therefore, the

primary and scattered light each pass through Glan–Thomson polarizers, and the orientation of the second polarizer (= analyzer) in front of the light scattering detector has to be carefully adjusted to a crossed position (with respect to the first polarizer). The scattered light was detected with an optical fiber setup coupled to a Thorn–Emi photomultiplier B2FBK/RFI, and the correlation function was determined from the intensity fluctuations with an ALV5000 multi-τ hardware correlator. Intensity autocorrelation functions were measured at scattering angles 20° to 150°, corresponding to a q range between 5.65 e-3 nm^{-1} and 31.4 e-2 nm^{-1}.

(iii) Methods used for data analysis:
The normalized electric field correlation function for optically isotropic monodisperse rod-like particles is given as:

$$g_1(q,\tau) = S(q,\tau) = S_0(qL)e^{-q^2 D_T \tau} + S_1(qL)e^{-(q^2 D_T + 6D_R)\tau} \tag{5.17}$$

with D_T the translational selfdiffusion coefficient, D_R the rotational selfdiffusion coefficient, L the length of the rod, and S_0 and S_1 the amplitudes of a purely translational relaxation process and a secondary relaxation mode containing contributions both from translational and rotational diffusion. At scattering angles corresponding to a length scale $qL < 3$, only translation contributes to the relaxation process, whereas at larger q (or smaller length scales) rotational motion becomes increasingly important. Equation 5.17 is valid for experimental length scales up to $qL < 8$. In a depolarized dynamic light scattering (DDLS, vh-geometry) measurement, the purely translational particle motion ($S_0(qL)e^{-q^2 D_T \tau}$, see Eq. 5.17) does not contribute to the correlation function, and the detected amplitude correlation function therefore is given as:

$$g_{1,vh}(q,\tau) = S_1(qL)e^{-(q^2 D_T + 6D_R)\tau} = S_{vh}(qL)e^{-\Gamma\tau} = S_{vh}(qL)e^{-(q^2 D_T + 6D_R)\tau} \tag{5.18}$$

Plotting the decay rate Γ versus q^2, the rotational diffusion coefficient D_R can be determined from the intercept $q \to 0$, and the translational diffusion coefficient D_T from the slope of the curve.

Using Broersma's relationship [5.4], in case of rigid noninteracting rods of aspect ratio $L/d > 5$ the diffusion coefficients are given as (see example C):

$$\begin{aligned} D_T &= \frac{kT}{3\pi\eta_0 L}\left(\ln\left(\frac{2L}{d}\right) - \xi\right) \\ D_R &= \frac{3kT}{\pi\eta_0 L^3}\left(\ln\left(\frac{2L}{d}\right) - 0.5(\gamma_\parallel + \gamma_\perp)\right) \end{aligned} \tag{5.19}$$

To express the parameters $\xi, \gamma_\parallel, \gamma_\perp$ as functions of the sample characteristics L and d, the authors have used the expressions derived by Tirado et al. [5.5], which are valid for rigid rods with aspect ratios L/d between 2 and 30. Equations 5.18, 5.19 provide the basis to determine L and d of rigid rods from the experimental results obtained in a DDLS measurement. The decay rate Γ (see Eq. 5.18) was determined by a Cumulant fit to the amplitude correlation function $g_{1,vh}(q,\tau)$ measured in vh-geometry. In addition, Lehner et al. determined the apparent diffusion coefficient $D_{eff}(q)$ from Cumulant fits of $g_{1,vv}(q,\tau)$, measured at various scattering angles in nonpolarized standard dynamic light scattering experiments in vv-geometry. Note that according to Eq. 5.17 $D_{eff}(q)$ should contain contributions from purely translational diffusion and from the secondary relaxation process, which consists both of translational and rotational diffusion components. Importantly, for anisotropic particles like TMV the translational diffusion coefficient D_T only provides a particle mobility average. D_T can be divided into two contributions: $D_{\parallel}$, the (larger) translational diffusion coefficient corresponding to a faster particle motion in direction of the rod axis, and $D_{\perp}$, the (smaller) translational diffusion coefficient corresponding to a slower particle motion perpendicular to the rod axis. The difference of these two particle mobility components $\Delta = D_{\parallel} - D_{\perp}$ is called the anisotropy of the average diffusion coefficient D_T.

(iv) Experimental results:
Figure 5.11 shows the decay rates obtained from DDLS measurements of a TMV solution in pure water, plotted versus q^2.
By linear regression analysis of the data plotted in Fig. 5.11, according to Eq. 5.18 the authors determined $6D_R = 1{,}798 \pm 60\,s^{-1}$ as the intercept and $D_T = (3.61 \pm 0.10) \cdot 10^{-12}\,m^2 s^{-1}$ as the slope of the linear curve. Compared with literature values, they found their result for the rotational mobility in agreement

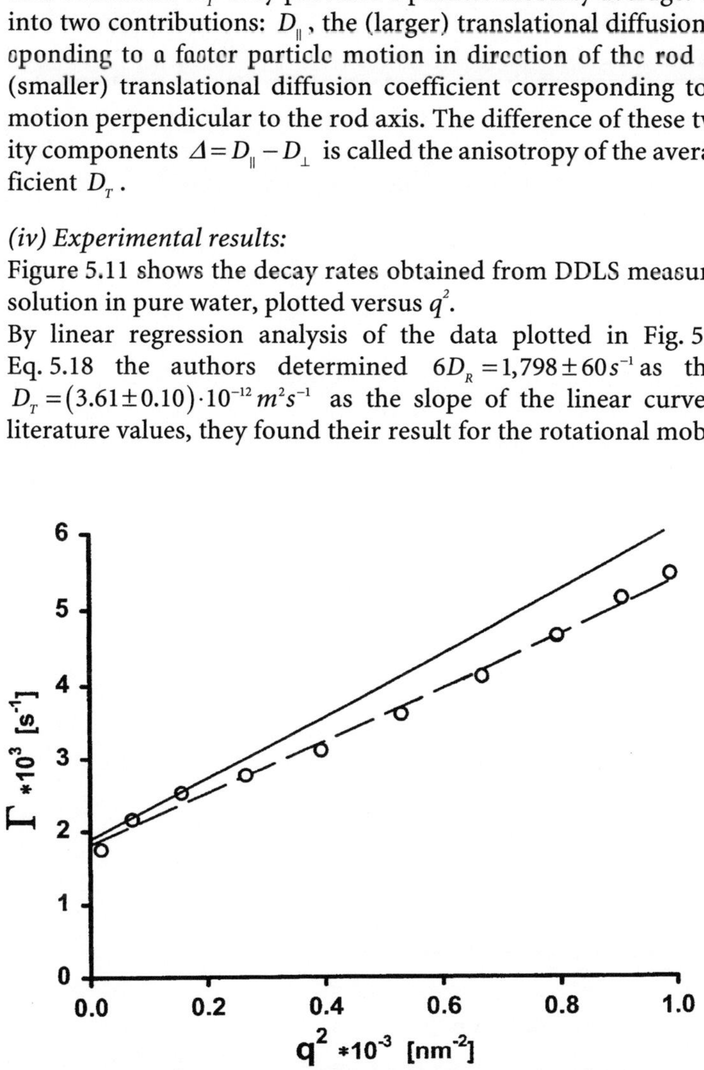

Fig. 5.11. q^2-dependence of decay rate obtained from DDLS experiments of TMV solutions. Reprinted with permission from D. Lehner, H. Lindner and O. Glatter, Langumir 16, 1689–1695, 2000, Copyright 2000 American Chemical Society (ref. [5.8]).

but their translational diffusion coefficient as about 20% too small. The expected value corresponding to a larger slope is shown by the solid line in Fig. 5.11.

Attributing this deviation to charge interactions, Lehner et al. decided to screen these effects by studying a TMV solution with added salt (2 e-3 M NaCl). Figure 5.12 shows a comparison of decay rates from DDLS measurements of TMV samples with and without added salt, plotted vs. q^2 .

The data show an increase of the slope of the curve and therefore in translational mobility, if salt screens the Coulomb interactions between the TMV particles in aqueous solution, while the rotational mobility corresponding to the intercept of the linear curves remains unchanged. The results obtained from fitting the data measured for the added salt solution are $D_R = 299 \pm 9 s^{-1}$ and $D_T = (4.05 \pm 0.09) \cdot 10^{-12} m^2 s^{-1}$. Calculations of the theoretically expected diffusion coefficients according to the expressions of Broersma, assuming $L = 300$ nm and $d = 18$ nm, yield $D_R = 295 s^{-1}$ and $D_T = 4.0 \cdot 10^{-12} m^2 s^{-1}$. Both values are in excellent agreement with the experimental result, illustrating the potential of the DDLS experimental technique.

The authors also used standard nonpolarized dynamic light scattering to investigate the same TMV solution with added salt. The resulting effective diffusion coefficients determined by a Cumulant analysis of $g_1(q,\tau)$, measured in vv-geometry, are shown in Fig. 5.13.

Using the rotational diffusion coefficient obtained from the DDLS experiment as input and leaving the translational diffusion coefficient D_T and its anisotropy $\Delta = D_{\|} - D_{\perp}$ as free fitting parameters, the best fit (solid line in

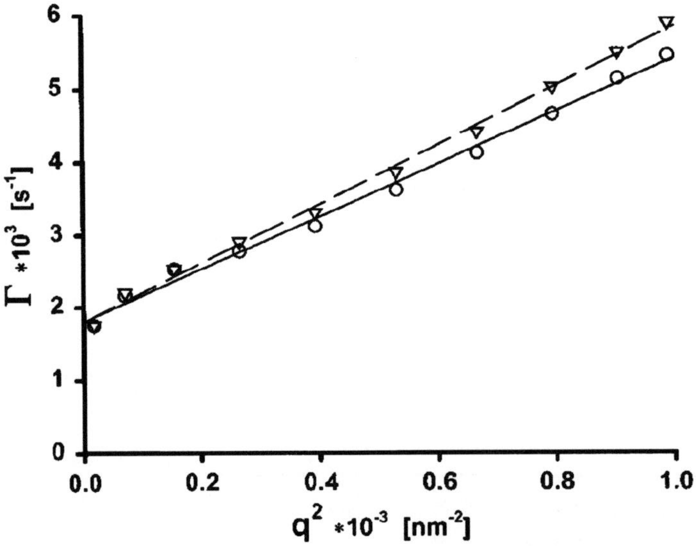

Fig. 5.12. q^2-dependence of decay rates from DDLS experiments of TMV solutions with (*triangles*) and without (*circles*) added salt. Reprinted with permission from D. Lehner, H. Lindner and O. Glatter, Langumir 16, 1689–1695, 2000, Copyright 2000 American Chemical Society (ref. [5.8]).

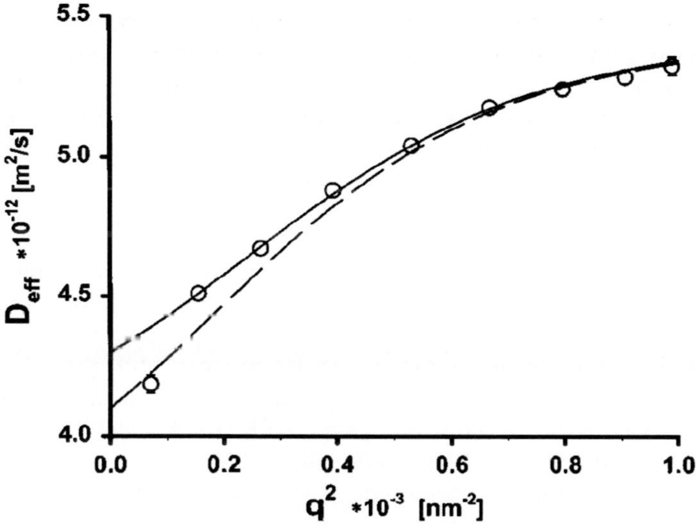

Fig. 5.13. q^2-dependence of effective diffusion coefficients determined from DLS experiments of TMV solutions with added salt. Reprinted with permission from D. Lehner, H. Lindner and O. Glatter, Langumir 16, 1689–1695, 2000, Copyright 2000 American Chemical Society (ref. [5.8]).

Fig. 5.13) corresponded to $D_T = 4.3 \cdot 10^{-12}\, m^2 s^{-1}$ and $\Delta = 1.76 \cdot 10^{-12}\, m^2 s^{-1} = 0.41 \cdot D_T$. Inserting both D_R and D_T as input parameters and only leaving the anisotropy Δ as a free-fitting parameter, another best fit (dashed line in Fig. 5.13) was obtained with $\Delta = 1.03 \cdot 10^{-12}\, m^2 s^{-1} = 0.25 \cdot D_T$. This later value is much smaller than the value $\Delta \approx 1.8 \cdot 10^{-12}\, m^2 s^{-1}$ theoretically expected for the TMV from hydrodynamic considerations, which corresponds well to the result of the first fit. Here, the authors state that the first data point measured at a very small scattering angle of 20° might suffer from dust impurities, explaining its strong deviation from the first fitting curve (see Fig. 5.13).

(v) Conclusions:
We have seen here an example illustrating the potential of the depolarized dynamic light scattering technique in analyzing both rotational and translational particle mobility of anisotropic particles. Also, as in the previous example, adding salt to the sample has proven to be crucial if particle characterization of charged particles in aqueous solution is desired.

Example E (Ref. [5.9]):
This last example for dynamic light scattering of rod-like particles shows the importance of including stabilization layers, present in most colloidal systems, in the data analysis. The system studied here consisted of colloidal gold rods stabilized in solution with a polymer layer of adsorbed poly(vinylpyrrolidone) chains. This sample also addresses another important issue, namely light ab-

sorption by the sample during the light scattering measurement, which should be avoided since it interferes with the sample characterization due to unwanted side effects such as sample heating, beam expansion, convection, etc. Since exploration of these effects by dynamic light scattering is an interesting and challenging subject itself, the last two examples F and G presented in this Chap. 5.1 on DLS will deal with this unusual topic.

(i) Samples and sample treatment:
Colloidal gold nanoparticles with aspect ratios in the range $2.5 < L/d < 49$ were synthesized by electrodeposition of gold in nanopores of anodized alumina serving as a template. The obtained particles were dispersed in an aqueous medium by dissolving the membrane, and subsequent release of the substrate while the polymer poly(vinylpyrrolidone) (PVP from Fluka, M = 40,000 g/Mol) was added as a stabilizer. Spherical gold nanoparticles were prepared by the standard citrate method, and again PVP was added to obtain PVP-stabilized nanospheres as reference particles for the light scattering experiments. DLS measurements were conducted in very dilute solutions of concentrations up to 40 mg/L so interparticle interactions leading to a deviation from Stokes–Einstein–behavior can be neglected. The sample characteristics determined from transmission electron microscopy (TEM) are summarized in Table 5.4.

(ii) Light scattering setup:
DLS measurements were performed with a krypton ion laser (Spectra Physics series 2020) operating at a wavelength of 647.1 nm (red light). This wavelength is far from the absorption maximum of the gold nanoparticles ($\lambda \approx 500$ nm), so

Table 5.4. Sample characteristics of gold nanoparticles as determined by TEM, with σ_L and σ_D the standard deviations. Reprinted with permission from B.M.I. van der Zande, J.K.G. Dhont, M.R. Bohmer and A.P. Philipse, Langumir 16, 459–464, 2000, Copyright 2000 American Chemical Society (ref. [5.9]).

System	Length L [nm]	σ_L [nm]	Diameter d [nm]	σ_D [nm]	L/d
Sphere18	18	5	–	–	–
Sphere15	15	3	–	–	–
Rod2.6a	47	17	18	3	2.6
Rod2.6b	39	10	15	3	2.6
Rod8.9	146	37	17	3	8.9
Rod12.6	189	24	15	3	12.6
Rod14	283	22	20	3	14.0
Rod17.2	259	60	15	3	17.2
Rod17.4	279	68	16	3	17.4
Rod39	660	20	17	3	39.0
Rod49	729	–	15	3	49.0

unwanted side effects caused by light absorption and corresponding local heating of the sample can be neglected (see also examples G and H for comparison). Normalized intensity autocorrelation functions were determined using a Malvern Multi 8 7032 CE 128-point correlator. Scattering angles ranged from 30° to 120°. Measurements were performed both in the vv-mode and the vh-mode (depolarized dynamic light scattering DDLS). The technical details of these two different experimental procedures have been described in the previous example.

(iii) Methods used for data analysis:
Intensity autocorrelation functions were fitted to single exponential decays, including a second Cumulant to account for particle size polydispersity:

$$g_2(q,\tau) = y = \alpha + \beta \exp\left(-b(q)\tau + c(q)\tau^2\right) \tag{5.20}$$

with fit parameters α and β. The decay rate b depends on the experimental method. In vv-mode, it is given as $b(q)=2D_T q^2$. In depolarized dynamic light scattering (vh-mode), the decay rate is given as $b(q)=2D_T q^2+12D_R$ (see Eq. 5.18). $c(q)$ is the second Cumulant providing a measure for the particle size polydispersity. As in the preceding example D, D_T is the translational and D_R the rotational selfdiffusion coefficient. The equations for the decay rates $b(q)$ are only valid in case of optically isotropic particles and larger experimental length scales, that is $qL<5$, in which case either no rotation is detected (vv-mode) or rotation-translation-coupling plays no role (vh-mode).

(iv) Experimental results:
The assumption of $qL<5$ is not justified for all gold rod species listed in Table 5.4, as shown in Fig. 5.14 presenting a comparison of the q-dependence of decay rates measured in vv and vh geometry for very small rods (sample ROD8.9) and very large rods (sample ROD39), respectively.

For the small rods, both the vv- and vh-based DLS measurements lead to the linear behavior predicted for $b(q)$ by the equations given in paragraph (iii). In contrast, the larger rods show a systematic deviation of the data points from the linear fit at larger q^2. This is to be expected, since the experimental scattering vector regime for these rods reaches a value $qL \approx 9$, corresponding to an experimental length scale regime where rotation-translation coupling leads to a nonlinear behavior. To apply the linear relations for the decay rate $b(q)$, van der Zande et al. have restricted the fitting range for these larger rods to values $q^2 < 2\cdot 10^{14}\, m^{-2}$, as shown in Fig. 5.14.

The translational diffusion coefficient is determined from the slope of these linear fits, and the rotational diffusion coefficient from the intercept ($q^2 \to 0$) of the data measured in vh geometry.

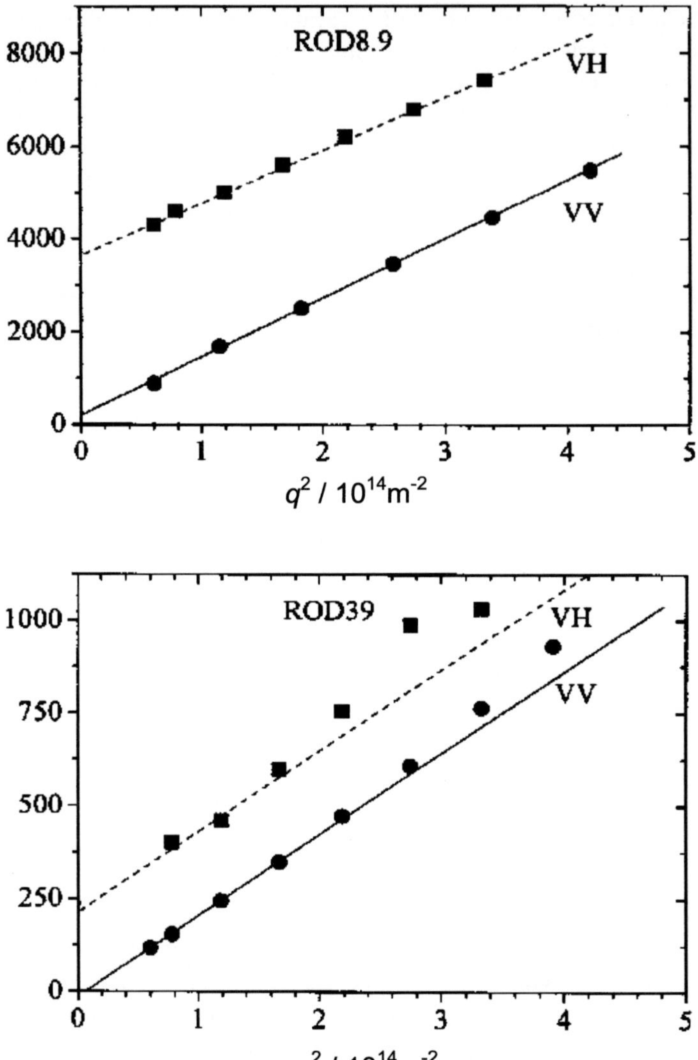

Fig. 5.14. q^2-dependence of DLS decay rates (in s^{-1}) for small (*top*) and large gold rods (*bottom*), measured in vv and vh geometry. Reprinted with permission from B.M.I. van der Zande, J.K.G. Dhont, M.R. Bohmer and A.P. Philipse, Langumir 16, 459–464, 2000, Copyright 2000 American Chemical Society (ref. [5.9]).

All experimental results have been summarized in Table 5.5. For comparison, diffusion coefficients have been calculated according to predictions of Tirado and de la Torre [5.5], using as input parameters length and diameter of the rods as determined from transmission electron microscopy:

Table 5.5. Experimental diffusion coefficients from light scattering and calculated values, using the TEM-based sample characteristics as input. Reprinted with permission from B.M.I. van der Zande, J.K.G. Dhont, M.R. Bohmer and A.P. Philipse,Langumir 16, 459–464, 2000, Copyright 2000 American Chemical Society (ref. [5.9]).

System	$10^{-12}\ D_T$ exp $[m^2 s^{-1}]$	$10^{-12}\ D_T$ cal $[m^2 s^{-1}]$	D_R exp $[s^{-1}]$	D_R cal $[s^{-1}]$
Rod8.9	6.0	8.4	306	2,238
Rod12.6	4.9	7.4	281	1,258
Rod14	2.9	5.2	66	396
Rod17.2	4.0	6.0	177	563
Rod17.4	3.5	5.6	175	452
Rod39	1.2	2.9	14	46
Rod49	0.7	2.8		30

$$
\begin{aligned}
D_T &= \frac{kT}{3\pi\eta L}\left[\ln\left(\frac{L}{d}\right)+0.312+0.565\frac{d}{L}+0.100\left(\frac{d}{L}\right)^2\right] \\
D_R &= \frac{3kT}{\pi\eta L^3}\left[\ln\left(\frac{L}{d}\right)-0.662+0.917\frac{d}{L}-0.050\left(\frac{d}{L}\right)^2\right]
\end{aligned}
\tag{5.21}
$$

These calculated values are also given in Table 5.5.

Obviously, there are strong deviations between experimental results determined by dynamic light scattering and diffusion coefficients calculated from the TEM results. All values determined by light scattering are systematically too small, which means the particle dynamics is slower than expected from the particle length and diameter determined by TEM. This discrepancy can be explained by considering that in solution, where the light scattering experiments took place, the particle surface was coated with a swollen polymer layer of PVP. This polymer layer, which becomes destroyed and therefore invisible in the TEM experiment, has to be added to the dimensions used in the calculations according to Eqs. 5.21, that is $L = L_{TEM} + 2\delta$ and $d = d_{TEM} + 2\delta$, with δ the layer thickness of the polymer coating which should be in the order of 5–20 nm. If a layer thickness of 10–15 nm is assumed, experimental data and theoretical calculations according to the authors are in good agreement.

(v) Conclusions:

The studies of anisotropic colloidal gold nanoparticles described above show that sometimes a combination of various experimental methods is needed to obtain a satisfactory sample characterization, as for example TEM and DLS in this case. Whereas the former allows one to accurately determine the sample topology, it may neglect certain features of the sample present in solution and not visible under the electron microscope, like a surface layer of organic surfactants or polymeric molecules around the scattering nanoparticles. Such surface coatings are often used in experimental practice to prepare stable colloidal nanoparticle dispersions.

Example F (Ref. [5.10]):
We have learned so far that dust particle contaminations within light scattering samples should be avoided at all costs due to their strong scattering intensity (typical size of dust particles $d \geq 1\mu m$, and scattered intensity $I \sim d^6$). In most cases, filtration, sometimes followed by centrifugation, provides the solution to this problem. However, for example, in case of larger sample particles or samples which adsorb easily at the filter membrane, filtration might not be an option. Here, example F may describe an alternative route: it shows that dust impurities and their contribution to the dynamic light scattering signal in principle can be treated analytically to allow particle sizing even for nonfiltered samples.

(i) Samples and sample treatment:
The procedure by Ruf described here was tested with aqueous solutions of a commercial surfactant, octaethylene glycol dodecyl monoether ($C_{12}E_8$). In aqueous solution this surfactant forms small spherical micelles of size <10 nm. The sample solution was prepared using distilled water from a Millipore filtration system and degassed. Sample concentrations were 30 mm. Dust was removed from the samples by filtration through 0.2-μm and 0.08-μm Nucleopore filters, and subsequent centrifugation for at least 10 min at 16,000 g using a Biofuge Pico table centrifuge from Heraeus. As deduced from the presence of a second slow process in the correlation function determined by dynamic light scattering, the samples purified in such way still contained a second class of particles about one order of magnitude larger than the micelles. These larger particles, assumed to be by-products from the synthesis of the surfactant, can be reduced in amount by filtration and centrifugation. Here, they will serve as "model dust", where the amount of "dust" can be adjusted by the total duration of centrifugation.

(ii) Light scattering setup:
Light scattered at an angle of 90° was detected through a few-mode fiber (SM-fiber 1064-FC with collimator from Linos, Göttingen) with an avalanche photo diode (SPCM-AQ-131-FL, EG&G Canada). Correlation functions $g_2(\tau)$ were calculated from the intensity fluctuations using two Brookhaven 9000AT correlators with linear and nonlinear correlation time spacing, respectively. To avoid afterpulsing from the avalanche photodiode, the first two correlation time channels were discarded during data analysis.

(iii) Methods used for data analysis:
The method for including dust contribution to dynamic light scattering signals is based on extending the usual scheme by two additional parameters, an offset in the amplitude autocorrelation function $g_1(\tau)$ and an offset in the intensity

autocorrelation function $g_2(\tau)$. Normalized intensity and normalized field autocorrelation function are then related according to the following quadratic equation:

$$g_2{}^* = g_2(\tau) - 1 = a|g_1(\tau)|^2 + b|g_1(\tau)| + c \tag{5.22}$$

In case that no dust is present in the sample, the equation simplifies to $b = c = 0$ and the well-known Siegert relation (Eq. 1.48 in Chap. 1.3) is obtained. So-called Gaussian dust causes a randomly fluctuating contribution to the scattered intensity, and therefore $b = 2(ac)^{0.5}$. Stationary dust leads to a constant contribution to the scattered intensity, and consequently $b \neq 0, c = 0$. Finally, so-called more-than-Gaussian dust corresponds to $b = 0, c \neq 0$. Accordingly, Gaussian dust produces an offset in the field correlation function while more-than-Gaussian dust produces an offset in the intensity correlation function.

An important parameter for the analysis of correlation functions in practice is the erroneous baseline $\hat{B}$, that is the value reached by the normalized intensity correlation function $g_2{}^*(\tau)$ at infinite correlation times, in respect to the exact baseline B, and the corresponding relative baseline error $\Delta B/\hat{B} = (\hat{B} - B)/\hat{B}$. Including this baseline error, the normalized intensity autocorrelation function can be expressed as:

$$g_2{}^*(\tau) = \left(1 - \frac{\Delta B}{\hat{B}}\right)\beta|g_1(\tau)|^2 - \frac{\Delta B}{\hat{B}} \tag{5.23}$$

with β the spatial coherence factor accounting for the number of speckles or coherence areas detected. Note that in case of dust-free samples ($\Delta B = 0$) and a coherence factor $\beta = 1$, $g_2{}^*(\tau) = |g_1(\tau)|^2$ as is to be expected. Equation 5.23 is analyzed for the limiting case of more-than-Gaussian dust, wherefore the former parameter c is identified as $c = -\Delta B/\hat{B}$. As a consequence, the baseline error has to be a negative value, or in other words the erroneous baseline must always be smaller than the true one, a prediction which is verified in experimental DLS data containing dust contributions. Introducing the offset B_c in the field correlation function $g_1(\tau)$ as the second parameter of interest, one can derive the following equation:

$$g_2{}^*(\tau) = \left(1 - \frac{\Delta B}{\hat{B}}\right)\left[\left(1 - \frac{B_c}{\beta^{0.5}}\right)\beta^{0.5}|g_{1,s}(\tau)| + B_c\right]^2 - \frac{\Delta B}{\hat{B}} \tag{5.24}$$

with $g_{1,s}(\tau)$ the idealized field autocorrelation function for the scattering particles of interest only, that is without any dust contribution. Correspondingly, the parameters from the phenomenological quadratic equation (Eq. 5.22) are identified as

$$
\begin{aligned}
a &= \left(1-\frac{\Delta B}{\hat{B}}\right)\left(1-\frac{B_c}{\beta^{0.5}}\right)^2 \beta \\
b &= 2\left(1-\frac{\Delta B}{\hat{B}}\right)\left(1-\frac{B_c}{\beta^{0.5}}\right)\beta^{0.5} B_c \\
c &= \left(1-\frac{\Delta B}{\hat{B}}\right)B_c^2 - \frac{\Delta B}{\hat{B}}
\end{aligned}
\tag{5.25}
$$

This set of equations provides a possibility to determine the "ideal" amplitude correlation function $g_{1,s}(\tau)$ from the measured intensity autocorrelation function $g_2(\tau)$, using baseline error $\Delta B/\hat{B}$, coherence factor β and offset B_c of the amplitude correlation function as input parameters. In detail, the amplitude correlation function without dust contribution can then be calculated as:

$$
\frac{g_2(\tau)+\dfrac{\Delta B}{\hat{B}}}{\left(1-\dfrac{\Delta B}{\hat{B}}\right)} = \left(\hat{g}_{1,\exp}(\tau)\right)^2 \text{ and } \left|g_{1,s}(\tau)\right| = \frac{1}{\beta^{0.5}} \frac{\hat{g}_{1,\exp}(\tau)-B_c}{\left(1-\dfrac{B_c}{\beta^{0.5}}\right)} \tag{5.26}
$$

with $\hat{g}_{1,\exp}(\tau)$ the experimental amplitude correlation function still suffering from dust contribution. First the experimental data are corrected with different pairs of the two parameters $\Delta B/\hat{B}$ and B_c to obtain the amplitude autocorrelation function of interest $g_{1,s}(\tau)$, which then is analyzed with CONTIN for the particle size distribution. For details how the fitting parameters are determined technically, which is beyond this book, the interested reader is referred to ref. [5.10]. To illustrate the success of this method, we will proceed here directly to some experimental results.

(iv) Experimental results:
Five different data sets of experimental intensity autocorrelation functions have been measured for the $C_{12}E_8$-sample to check the new procedure. Data sets 1–3 are determined from the same raw data measured for one given sample after filtration and only 10 min of centrifugation. These data sets therefore should suffer from a comparatively large "dust" contribution. The DLS measurements were frequently repeated using two different hardware correlators, one operating in linear mode with sampling time 2 μs and 80 channels, the other operating in so-called ratio mode with two different sampling times of 1 μs and 5 μs, totalling 206 channels. Covered delay times range from 2 to 160 μs for the linear correlator and from 1 μs to 4 ms for the nonlinear correlator. 720 runs or 720 single autocorrelation functions were measured in total. Data set no. 1 was calculated as an average over all 720 runs determined with the linear correlator only. For this reason, it is expected that data set 1 contains the largest contribution of dust to the signal. Data set no. 2 was obtained by omitting the 52 runs with mean count rates much higher than those of the other 668 runs from the linear correlator

runs, before calculating the average intensity correlation function $g_2(\tau)$, thereby reducing the contribution of dust. Data set no. 3 was determined in a way similar to data set no. 2, but this time considering the correlation functions determined with the nonlinear correlator and consequently extending the averaged correlation function to longer delay times. For the remaining two data sets, first the sample was further purified by additional centrifugation for another 30 min. Then, a data set of 720 correlation functions was measured with two linear correlators in a combined arrangement, both using a sampling time of 2 μs but different delay times of 2 μs and a multiple of 2 μs, respectively. Then, the correlation functions determined by the two linear correlators for a given run were spliced together to extend the total delay time range, using an intermediate time-scale overlap of 10–20 channels. Data set no.4 was obtained in this way from an average omitting 31 outliers with higher mean count rates. Finally, the sample was kept for about 24 h at temperature 50°C which led to an additional considerable reduction of "dust". The 960 measurements, performed with this sample and used for determining data set no.5 by averaging, showed only one outlier of increased count rate compared to the mean value, proving the successful removal of dust.

Field autocorrelation functions determined by the usual way, i. e., by subtracting 1 from the normalized intensity autocorrelation functions and extracting the square root, are shown in Fig. 5.15.

As expected, data set no.1 shows the largest indication of dust as indicated by the large offset in the correlation function.

Figure 5.16 shows the output of a CONTIN analysis of data set no.2 without additional corrections (left), and after reducing dust-contribution from the amplitude correlation function by determining suitable values for the parameters $\Delta B/\hat{B}$ and B_c, following the procedure described above.

Also shown in the figure are the residuals, which provide a measure for the quality of the CONTIN fits. Obviously, the new method leads to a much better quality of the CONTIN results as proven by the much smaller residuals, which also show a random statistical deviation of the fit from the experimental data in contrast to a systematic error for the analysis without additional corrections (Fig. 5.16, left).

In addition, the mean size of the micelles shifts from 2.4 nm to 3.3 nm, underlining the necessity of the corrections to obtain reliable values for the particle size. Parameters used for the calculation of corrected correlation functions, shown on the right hand side of Fig. 5.16, are (on average, for details see ref. [5.10]) $\Delta B/\hat{B} = -2 \cdot 10^{-4}$ and $B_c = 0.02$.

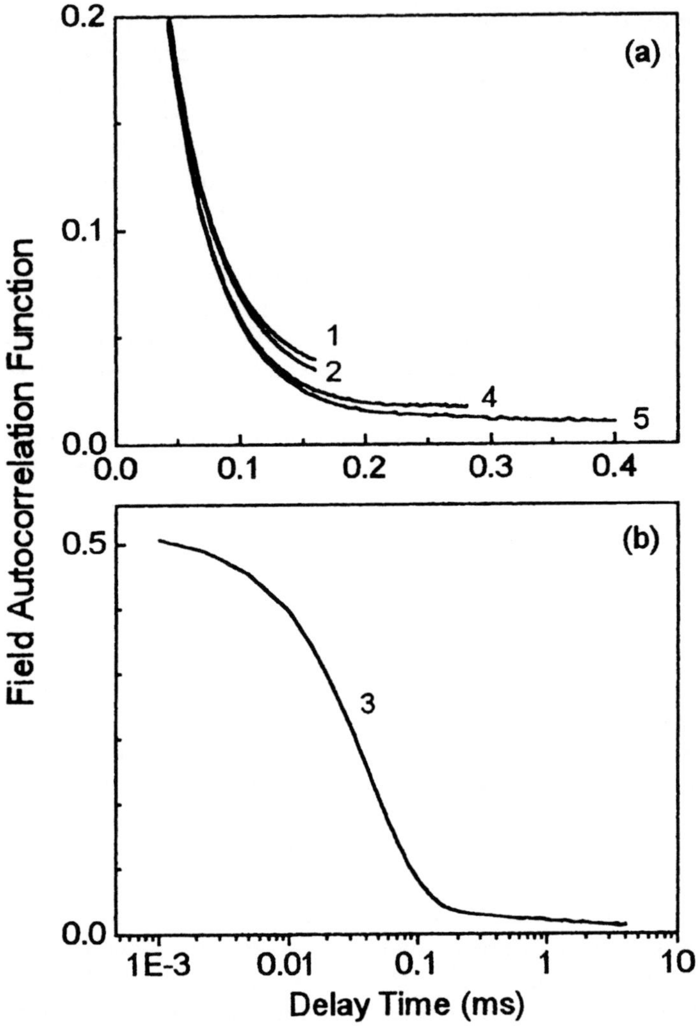

Fig. 5.15. Field autocorrelation functions for data sets 1–5 (see text for details). Reprinted with permission from H. Ruf, Langumir 18, 3804–3814, 2002, Copyright 2002 American Chemical Society (ref. [5.10]).

Parameters for correction of all five data sets are given in Table 5.6.

As expected, the parameters $\Delta B/\hat{B}$ and B_c decrease systematically with decreasing contribution of dust to the measured correlation functions (from sample 1 to sample 5). Finally, Ruf has shown that all data sets measured with the linear correlators (data sets no. 1, 2, 4, and 5) can be superimposed after removing the dust contribution, using the parameters given in the table for respectively calculating the corrected correlation functions $g_{1,s}(\tau)$.

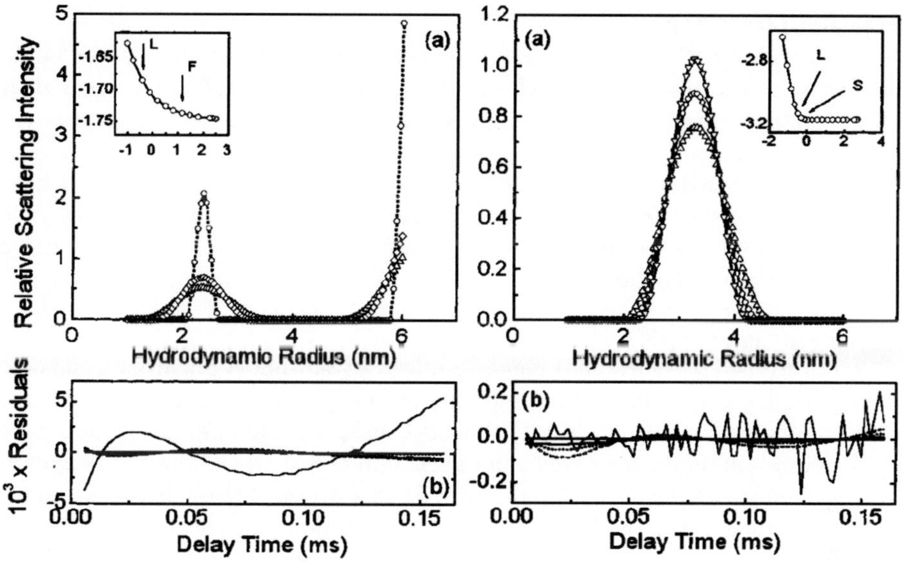

Fig. 5.16. CONTIN analysis of field autocorrelation functions for data set 2 without additional corrections (*left*), and with corrections for $\Delta B/\hat{B}$ and B_c (*right*). Reprinted with permission from H. Ruf, Langumir 18, 3804–3814, 2002, Copyright 2002 American Chemical Society (ref. [5.10]).

Table 5.6. DLS correlation function parameters $\Delta B/\hat{B}$ and B_c to treat a varying dust contribution for five data sets (see text for details). Reprinted with permission from H. Ruf, Langumir 18, 3804–3814, 2002, Copyright 2002 American Chemical Society (ref. [5.10]).

Data set no.	$\Delta B/\hat{B}$	B_c	B
1	-4.5×10^{-4}	2.1×10^{-2}	8.3×10^{9}
2	-7.5×10^{-5}	2.1×10^{-2}	7.7×10^{9}
3	-1.1×10^{-4}	–	–
4	-2.1×10^{-4}	8.1×10^{-3}	7.9×10^{9}
5	-4.0×10^{-5}	9.5×10^{-3}	10.6×10^{9}

(v) Conclusions:

This example did not only illustrate how dust impurities in dynamic light scattering experiments may lead to strong errors in particle sizing. It also presented a valuable approach how such unwanted dust contributions to the correlation function could be eliminated in case physical removal of the dust from the sample itself by filtration, etc. is not an option.

Example G (Ref. [5.11]):
This example briefly reviews previous experiments of the author of this book. Intending to use gold nanoparticles as tracers to study selfdiffusion in colloidal dispersions by standard dynamic light scattering, we encountered light absorption and local sample heating as unwanted side effects. Whereas these effects made an accurate determination of the single particle diffusion by dynamic light scattering, and therefore also accurate particle sizing, very difficult, they gave rise to an interesting approach towards quantification of convective flux using DLS, as will be shown here.

(i) Samples and sample treatment:
Samples studied were dilute toluene solutions of self-made polyorganosiloxane nanoparticles of diameter 75 nm (concentration 0.2 wt %), each containing one single tiny gold cluster. These gold clusters caused a broad light absorption peak of the sample solutions centered at a wavelength of 520 nm. At the wavelength of the incident laser light (514 nm), the specific absorption of the gold particle solutions investigated was 1.10 cm^{-1}. The sample solutions had been purified from dust by filtration through 5.0-μm pore size Millipore filters into cylindrical Suprasil glass light scattering cuvettes of 10 mm outer diameter. The cuvettes themselves had been cleaned before with refluxing acetone vapour using a home-built distillation setup.

(ii) Light scattering setup:
The DLS setup was a standard ALV goniometer setup, using an argon ion laser operating at wavelength 514 nm and laser power between 50 and 500 mW as the light source. Scattered light was detected with a single-mode fiber detector, and the time intensity autocorrelation functions $g_2(q,\tau)$ were calculated using an ALV5000 hardware correlator. During our measurements the sample was kept in a toluene bath at 20°C to keep the average sample temperature constant and to suppress reflections from the glass-air-interface of the sample cell. It should be stressed that this was a standard so-called homodyne DLS experiment, where the intensity autocorrelation function is measured and the amplitude autocorrelation function, which cannot be measured directly, could be calculated from the experimental data using the Siegert relation (Chap. 1.3, Eq. 1.48).

(iii) Methods used for data analysis:
In case scattering particles investigated in a DLS experiment exhibit both random translational diffusion and regular convective flow, the amplitude correlation function is given as a combination of these two relaxation processes according to:

$$g_1(q,\tau) = \exp\left(i\vec{q}\vec{v}\tau\right) \cdot \exp\left(-Dq^2\tau\right) \tag{5.27}$$

with D the selfdiffusion coefficient and $\vec{v}$ the velocity vector describing the regular particle motion. In a homodyne DLS experiment, the detected intensity correlation function is given as:

$$
\begin{aligned}
g_2(q,\tau)-1 &= g_1(q,\tau)\cdot g_1(q,\tau)^* = \\
&= \exp\left(i\vec{q}\vec{v}\tau\right)\cdot\exp\left(-Dq^2\tau\right)\cdot\exp\left(-i\vec{q}\vec{v}\tau\right)\cdot\exp\left(-Dq^2\tau\right) = \\
&= \exp\left(-2Dq^2\tau\right)
\end{aligned}
\tag{5.28}
$$

Here, the complex exponentials cancel and there seems to be no possibility to detect the relaxation contribution of the convection to the measured correlation function in a homodyne light scattering experiment.

(iv) Experimental results:
Light absorption of incident laser light by the sample lead to local heating which, as a consequence of the corresponding change in local solvent density due to thermal expansion, created a regular convective flow pattern. This flow pattern had been verified by measuring the local temperature within the irradiated sample with a tiny platinum temperature sensor. Since the average temperature of the sample was kept constant by the surrounding toluene bath, a stable flow pattern as deduced from the measured temperature profile was maintained (see Fig. 5.17).

Obviously, the flow pattern shown in Fig. 5.17 contains not only a single regular flow direction $\vec{v}$. Instead, two opposing fluxes $+\vec{v}$ and $-\vec{v}$ are located in the scattering plane, and therefore are detectable by the light scattering experiment, as illustrated in Fig. 5.17. As a consequence, the amplitude correlation function is given in a simplified picture (for more details, see ref. [5.11]), as:

$$
\begin{aligned}
g_1(q,\tau) &= \exp\left(i\vec{q}\vec{v}\tau\right)+\exp\left(-i\vec{q}\vec{v}\tau\right)\cdot\exp\left(-Dq^2\tau\right) - \\
&= 2\cos\left(\vec{q}\vec{v}\tau\right)\cdot\exp\left(-Dq^2\tau\right)
\end{aligned}
\tag{5.29}
$$

Equation 5.29 shows that velocity contributions to the measured intensity autocorrelation function $g_2(q,\tau)-1=g_1(q,\tau)\cdot g_1(q,\tau)^*$ will not cancel in our homodyne scattering experiment as in Eq. 5.28, but instead lead to regular oscillations with oscillation period given by $\vec{q}\cdot\vec{v}=qv\cos(\theta/2)=2\pi n_D\lambda^{-1}v\sin\theta$, with θ the scattering angle. The formula for the oscillation period is simply derived from the definition of the scattering vector and the scattering geometry shown in Fig. 5.17. In conclusion, we would expect to see regular oscillations both in the intensity and amplitude correlation functions of the homodyne DLS experiment due to two opposing velocity components. Importantly, the period of these oscillations should scale linearly with $\sin\theta$. Further, it should be possible to quantitatively determine the convective flux velocity from the slope of a plot of the oscillation frequency versus $\sin\theta$. Figure 5.18 shows one amplitude correlation function $|F(q,\tau)|=g_1(q,\tau)$ calculated from the intensity correlation function $g_2(q,\tau)$ using the Siegert relation, including a rough fit to the data derived from the model shown in Fig. 5.17.

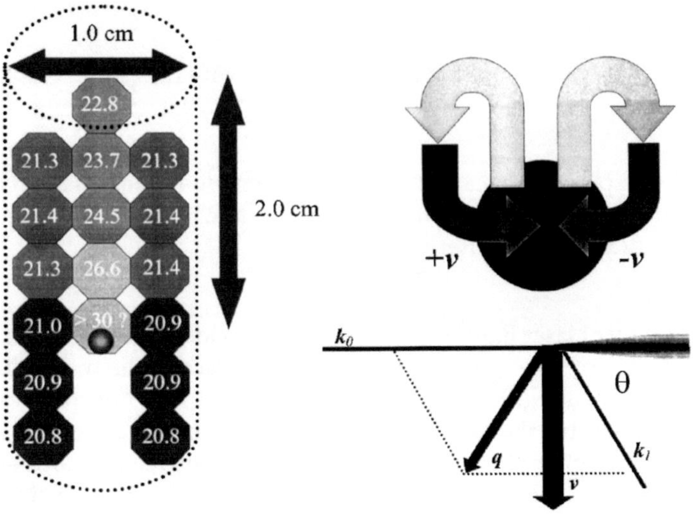

Fig. 5.17. Experimental temperature profile (*left*), according convective flow pattern and corresponding scattering geometry for a colloidal gold sample locally heated by laser light absorption. Reprinted with permission from W. Schaertl and C. Roos, Physical Review E 60, 2020–2028, 1999, Copyright (1999) by the American Physical Society (ref. [5.11]). http://prola.aps.org/abstract/PRE/v60/i2/p2020_1

Figure 5.18 shows also the dependence of the oscillation period, identified by the first minimum of the amplitude correlation function, on scattering angle.

The predicted linear behavior $\tau_{\min}^{-1} \sim \sin\theta$ is only roughly fulfilled by the experimental data, which can be explained simply as follows: local heating of the sample causes a modulation of the refractive index of the solution, leading to an expansion of the laser beam, a phenomenon which usually is called a "thermal lens". This effect causes a slight change in the scattering geometry, which has to be taken into account when deriving the equation for the amplitude correlation function (Eq. 5.29). The slightly changed scattering geometry and the correspondingly corrected data are shown in Fig. 5.19.

Using the beam expansion correction, the predicted linear behavior now is perfectly fulfilled. The expansion angle used as a fit parameter to correct the data was 5°, which was in excellent agreement with the beam expansion measured from the width of the transmitted beam at a certain distance behind the sample. From the slope of the curve shown in Fig. 5.19, a reasonable flow velocity $\nu = 1.6$ mm/s was calculated. Finally, it should be noted that ν depends on the light energy uptake per irradiation time, and therefore on sample concentration as well as on laser power. Also, in case of a solvent with smaller thermal expansion coefficient, for example for aqueous solutions of colloidal gold particles, no oscillations of the correlation function were observed in a dynamic light scattering experiment, indicating the absence of a regular convective motion. Here, the change in solvent density due to local heating of the sample solution was too small to create the regular convective flow pattern shown in Fig. 5.17.

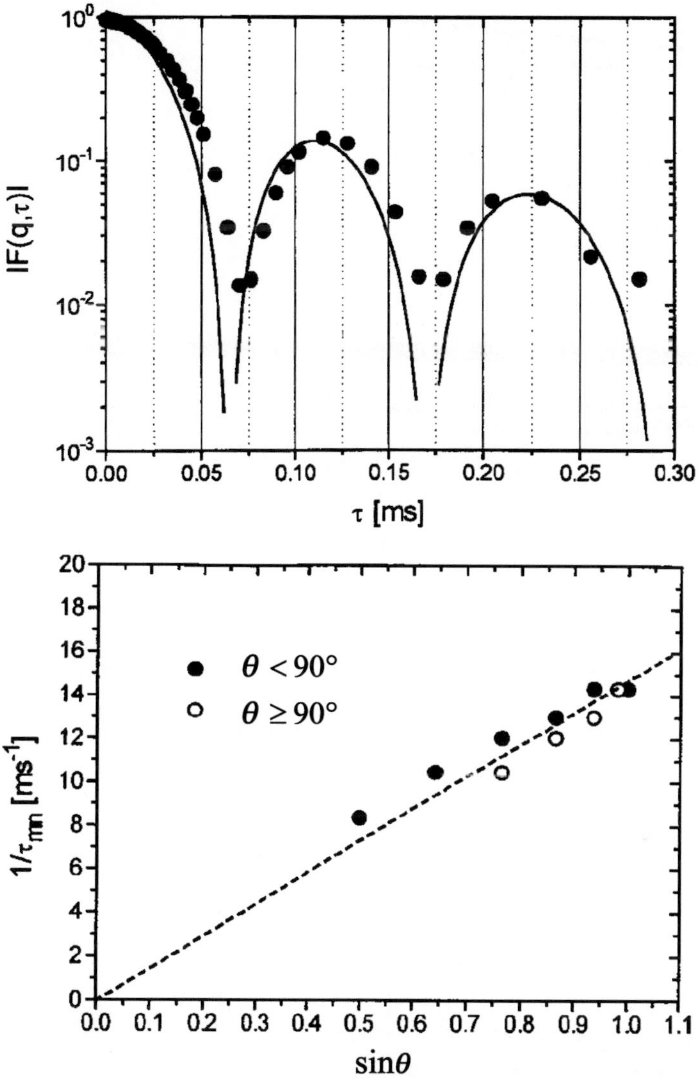

Fig. 5.18. Oscillating amplitude correlation function measured by DLS at laser power 500 mW and scattering angle 70°, including a fit using the simplified convection picture described in the text (*top*), and dependence of oscillation period on scattering angle (*bottom*). Reprinted with permission from W. Schaertl and C. Roos, Physical Review E 60, 2020–2028, 1999, Copyright (1999) by the American Physical Society (ref. [5.11]). http://prola.aps.org/abstract/PRE/v60/i2/p2020_1

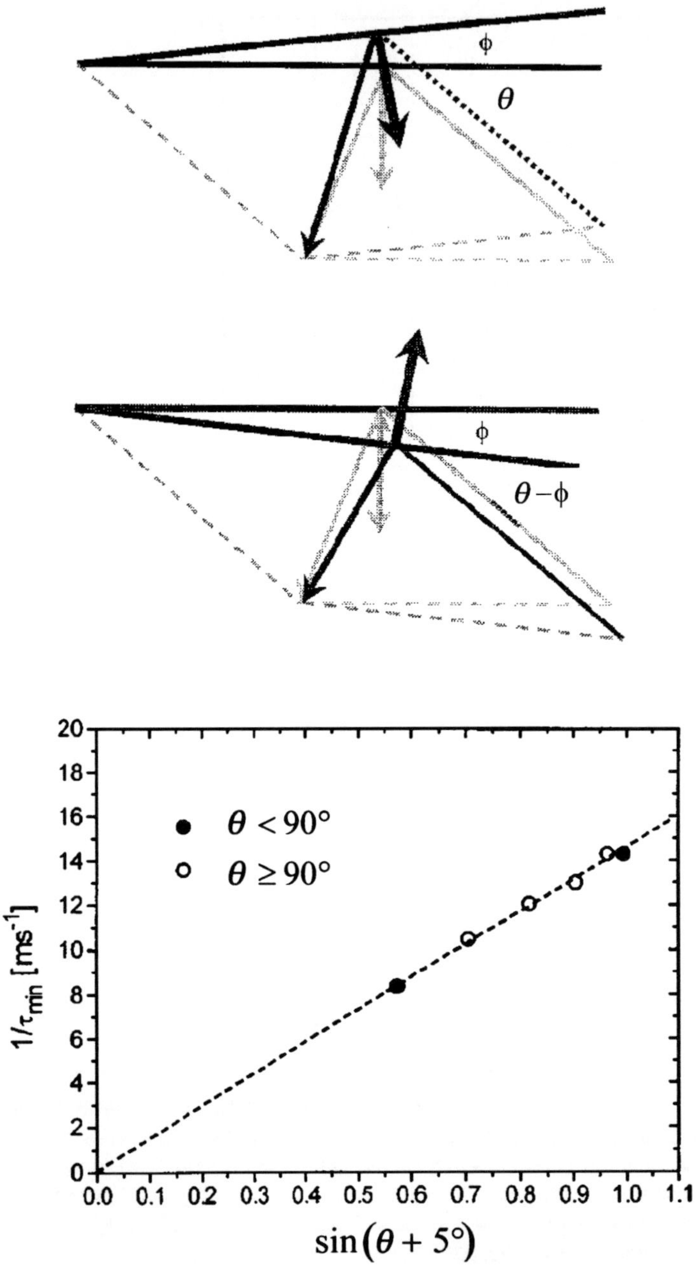

Fig. 5.19. Change in the scattering geometry due to the "thermal lens" effect (*top*), and accordingly corrected dependence of the correlation function oscillation period on scattering angle (see also **Fig. 5.18**, *bottom*). Reprinted with permission from W. Schaertl and C. Roos, Physical Review E 60, 2020–2028, 1999, Copyright (1999) by the American Physical Society (ref. [5.11]). http://prola.aps.org/abstract/PRE/v60/i2/p2020_1

(v) Conclusions:
Usually, light absorption is a phenomenon which should be avoided in dynamic light scattering since it prohibits accurate particle sizing by several effects: (1) local heating causes a change in particle diffusion due to viscosity changes (see Stokes–Einstein–equation) and (2) unusual dynamical processes may be created by the local heating, leading to changes in the correlation function itself. Example G nicely demonstrates how one can benefit from a disadvantage by developing a new model for data analysis of correlation functions detected in a homodyne dynamic light scattering experiment, allowing now the quantitative investigation of convective particle motion by dynamic light scattering.

Example H (Ref. [5.12]):
Example H also deals with the effect of light absorption on homodyne DLS measurements (such as example G). Although the signals look fairly similar in both cases, the authors Sehgal and Seery attribute the presence of oscillations in their correlation functions to so-called heterodyning caused by the heterogeneity or polydispersity of the investigated samples. Since this is an approach totally different from what just has been described in example G, it deserves a closer look.

(i) Samples and sample treatment:
A dilute m-cresol solution was prepared from the linear polymer polyanilin (PANI) in its dry powder form. Camphorsulfonic acid (CSA) and tetramethylammonium chloride (TMAC) were used to control the ionic strength of the solution due to their high solubility in m-cresol. Since TMAC is highly hygroscopic, it was dried in an oven at 200°C for 30 min prior to use. An aliquot of a 25-mg/mL stock made from this dry salt was added to a PANI solution to give the PANI/m-cresol/TMAC sample with polymer concentration $c_p = 0.0455$ mg/mL and $c_{TMAC} = 16.67$ mg/mL. This solution was filtered through a 1-µm pore size polytetrafluoroethylene (PTFE) 25-mm cartridge filter into a 13-mm cylindrical scattering cell which was cleaned from dust with condensing acetone vapour. During the experiment the temperature of the cell was controlled within a refractive index matching decalin bath.

(ii) Light scattering setup:
Dynamic light scattering experiments were conducted with a Brookhaven Instruments BI9000-AT autocorrelator with correlation times from 0.025 µs to 10 s. A water-cooled Coherent Innova 70-3 argon ion laser, operating at wavelength 488 nm or 514.5 nm, depending on the maximum amplitude of the observed correlation function, was used as the light source. The laser power ranged from 200 mW to 1 W. The sample was mounted on a Brookhaven BI200SM goniometer equipped with a decalin bath, enabling a scattering angular range between 20° and 150°. For lack of scattered intensity, the DLS measurements unfortunately were restricted to scattering angles between 20° and 60°.

(iii) Methods used for data analysis:
As stated in the preceding example G, usually a homodyne experiment should not be capable of detecting regular particle motion caused by light absorption. Nevertheless, the authors had detected anomalous periodic oscillation functions. Since their sample was a highly light absorbing solution of associating polymers, they assumed that a widely separated bimodal population of scattering particles could produce so-called heterodyning. This means that scattering from the larger fraction of particles, the associates or aggregates, serves as an incident beam which interferes with the scattering from the smaller sample fraction, thereby providing the phase information necessary for measuring the amplitude correlation function $g_1(q,\tau)$ directly. It should be noted that in a common heterodyne dynamic light scattering experiment this phase information is obtained by mixing the scattered light with the incident laser beam itself, carefully using a rather complicated optical adjustment. According to Sehgal and Seery, partial heterodyning caused the regular oscillations in their measured intensity autocorrelation functions. Let us review the formalism described in ref. [5.12] in more detail. The amplitude autocorrelation function for a mixture of two scattering particle size fractions each exhibiting diffusive and regular motion is given as:

$$g_1(\vec{q},\tau)=\langle N_1\rangle \exp\left[i(\vec{q}\vec{v}_1)\tau-q^2D_1\tau\right]+\langle N_2\rangle \exp\left[i(\vec{q}\vec{v}_2)\tau-q^2D_2\tau\right] \quad (5.30)$$

with $\langle N_1\rangle$ and $\langle N_2\rangle$ the amplitudes of the relaxation processes as determined by concentration and size of the respective particle fraction. To obtain the intensity autocorrelation function, $g_1(\vec{q},\tau)$ has to be multiplied with its complex conjugate:

$$\begin{aligned} g_2(\vec{q},\tau)&=g_1(\vec{q},\tau)\cdot g_1(\vec{q},\tau)^*=\\ &=\langle N_1+N_2\rangle^2+\langle N_1\rangle^2\exp(-2q^2D_1\tau)+\langle N_2\rangle^2\exp(-2q^2D_2\tau)+ \quad (5.31)\\ &\quad +2\langle N_1\rangle\langle N_2\rangle\exp(-2q^2(D_1+D_2)\tau)\cdot\cos(\vec{q}\cdot\Delta\vec{v}\cdot\tau) \end{aligned}$$

Equation 5.31 contains a cosine term representing oscillations caused by convection, and visible in the intensity autocorrelation function measured in the homodyne experiment. The heterodyning effect due to the presence of a second scattering species, according to the authors, enables quantification of the relative velocity difference between the two convectively flowing species ($\Delta\vec{v}=\vec{v}_2-\vec{v}_1$). Considering the convection model shown in Fig. 5.17 of example G, one may wonder why the two scattering particles in solution should exhibit different convective flow velocities $\vec{v}_2,\vec{v}_1$ instead of simply following the path directed by the convective macroscopic flux of the pure solvent. Therefore, one would assume $\Delta\vec{v}\approx 0$, making the scenario depicted in the previous example more likely than the partial heterodyning approach illustrated here in example H. Importantly, it should be pointed out that although the formalism of the two examples is totally different, the resulting equations in respect to the oscillatory term are identical, and it is impossible to distinguish which scenario is more plausible from the experimental dynamic light scattering data alone. The most important difference is that in example G the oscillation is given by the convective flux velocity itself, whereas here in example H it corresponds to the difference in flux velocity of species 1 and 2.

Including the thermal lens effect or beam expansion, Sehgal and Seery derived a modified expression for the scattering vector in comparison to the usual version:

$$\left|\vec{q}\right|_{MOD}^{2}=\left(\frac{4\pi n_D}{\lambda}\right)^2\left[\frac{1}{2}\left(1-\left(\cos\theta\cos^2\frac{\Delta}{2}\right)\right)\right] \tag{5.32}$$

instead of

$$\left|\vec{q}\right|^2=\left(\frac{4\pi n_D}{\lambda}\right)^2\left[\frac{1}{2}(1-\cos\theta)\right]=\left(\frac{4\pi n}{\lambda}\right)^2\sin^2\left(\frac{\theta}{2}\right) \tag{5.33}$$

Here, Δ is the beam expansion angle.

(iv) Experimental results:
Figure 5.20 shows the intensity autocorrelation functions measured at different scattering angles.

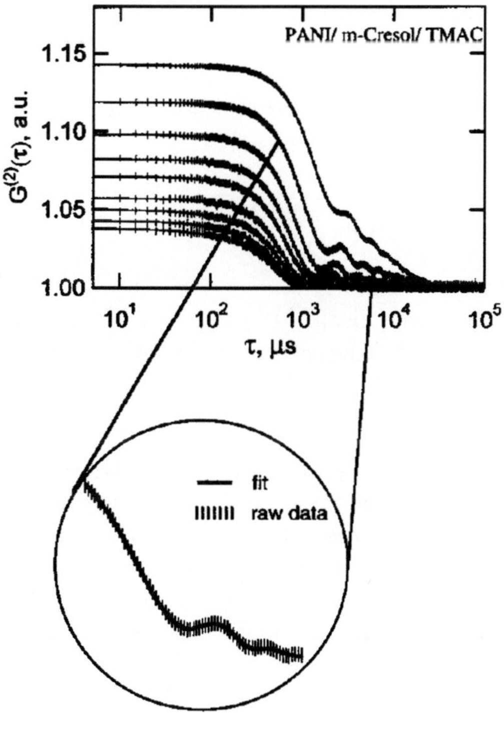

Fig. 5.20. Oscillating intensity correlation functions measured by DLS for scattering angles from 20° to 60° (from *top* to *bottom*). Reprinted with permission from A. Sehgal and T.A.P. Seery, Macromolecules 32, 7807–7814, 1999, Copyright 1999 American Chemical Society (ref. [5.12]).

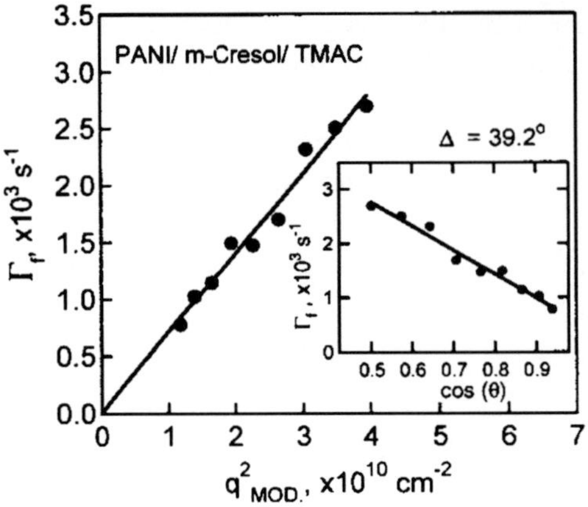

Fig. 5.21. DLS decay rate of the fast diffusive process, plotted vs. modified scattering vector and scattering angle (*inset*). Reprinted with permission from A. Sehgal and T.A.P. Seery, Macromolecules 32, 7807–7814, 1999, Copyright 1999 American Chemical Society (ref. [5.12]).

The data have been fitted using Eq. 5.31, showing reasonable agreement of experimental data and fit, although especially at higher oscillation periods the oscillations are not properly reproduced (see inset in Fig. 5.20).

Figure 5.21 shows the q-dependence of the decay rate of the fast diffusion mode D_1 as determined from the fitting functions shown in Fig. 5.20. In addition, the authors have also determined the diffusion coefficient and the expansion angle Δ from a plot of the decay rate $\Gamma = 2q^2 D_1$ (see Eq. 5.31) vs. $\cos\theta$, as shown in the inset of Fig. 5.21.

The value $D_1 = 3.5 \cdot 10^{-8}\, cm^2 s^{-1}$ seems to be reasonable for Stokes–Einstein diffusion of a polymer chain in a molecular solvent, whereas the beam expansion $\Delta = 39.2°$ seems to be comparatively large.

Finally, in analogy to example G, the velocity difference component Δv_z lying within the scattering plane has been determined from a plot of the oscillation period vs. $\cos\theta$ according to Eq. 5.34:

$$\frac{1}{\tau_V} = \Gamma_V = \frac{\Delta v_z \cdot n_D}{2\lambda} \cdot \left(\frac{\cos^2 \Delta}{1 - \cos \Delta} - 2\cos\theta \right) \tag{5.34}$$

These results are shown in Fig. 5.22.

From the slope of the curve shown in the inset of Fig. 5.22, the authors obtained the value $\Delta v_z = 0.0392\, cm \cdot s^{-1}$ which is comparable in magnitude to the convective flux velocity presented in example G.

(v) Conclusions:

This example shows how so-called heterodyning, either caused by a bimodal size distribution of the scattering particles themselves or willingly by addition of a second scattering species to the sample, might lead to oscillations in intensity autocorrelation functions from assumingly homodyne dynamic light scattering measurements. It remains an open question which scenario (example G or H) is the correct one to explain the oscillations found in homodyne DLS measurements of light absorbing samples.

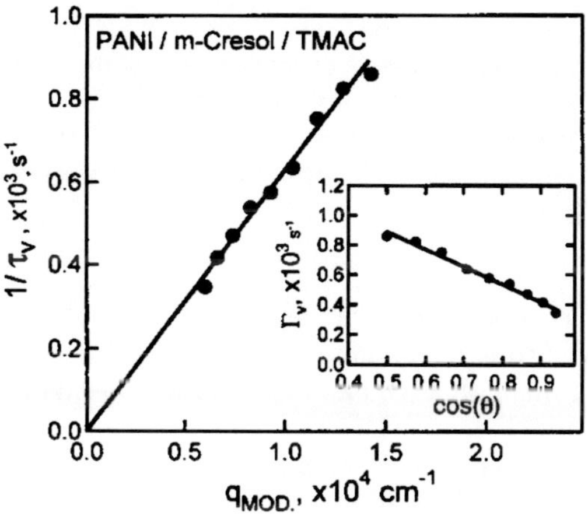

Fig. 5.22. Oscillation period and velocity decay rate of DLS correlation functions, plotted vs. modified scattering vector and scattering angle (*inset*). Reprinted with permission from A. Sehgal and T.A.P. Seery, Macromolecules 32, 7807–7814, 1999, Copyright 1999 American Chemical Society (ref. [5.12]).

5.2 Static Light Scattering

In this section, I will present several representative examples from the recent literature for the experimental practice of static light scattering and, in some cases, a clever combination of static and dynamic light scattering measurements. Beginning with linear homopolymer chains, copolymer chains and copolymer micelles, we will move on to a thorough investigation of branched polymers in solution by one of the leading experts in the field of light scattering from polymer solutions, professor Burchard from Freiburg University, Germany. Next, we will describe the investigation of the structure of small and large vesicles by light scattering. Vesicular topologies have become more and more important due to their potential for encapsulation and controlled release. Other important samples studied in light scattering experiments, which are treated here, are semiflexible and stiff cylindrical nanoparticles. Chapter 5.2 will conclude with an example of the coil-globule transition in aqueous solution of the important thermo responsive polymer poly-N-isopropylacrylamide (PNIPAM), and finally an example for the investigation of the formation of fractal aggregates by salt-driven aggregation of spherical nanoparticles. Note that the general practical limitations of the light scattering technique already discussed in Chaps. 4 and 5.1 also hold for static light scattering: dust, light absorption, and particle interactions giving rise to a structure factor contribution, the latter especially important for charged systems in seemingly dilute aqueous solutions, should be avoided by all means to allow for clean single particle characterization. Therefore, these limitations will not further be considered in detail in this chapter.

Example A (Ref. [5.13]):
Our first example shows the experimental standard procedure for static light scattering (SLS) from linear polymer chains in dilute solution. What is nonstandard here is the unusually high sample temperature >200°C which affords a special home-built apparatus both for sample preparation (= sample filtration) and the light scattering experiment itself.

(i) Samples and sample treatment:
Commercially available diisobutyl adipate, which was used here as a polymer solvent, was first purified by passage through a column of activated silica gel to achieve a final purity of 99.5%. Commercial polymer samples of an alternating copolymer of ethylene and tetrafluoroethylene (PETFE) were used as received. Solutions of different concentrations were prepared from these two components at 250°C. For dissolution and subsequent purification by filtration, the authors have designed a special high temperature apparatus.

(ii) Light scattering setup:
For their light scattering experiment at high temperature, Chu and Wu also constructed a high-temperature setup capable of maintaining easily a temperature of 200–230°C with temperature fluctuations less than 0.2°C. Details of this double-thermostat-setup are described in ref. [5.13]. The detector consisted of two detector arms located 90° apart which could be rotated, each equipped with a lens defining the scattering volume to a scattering angle accuracy of $\delta\theta \leq \pm 0.1°$. The scattered light was transferred to a photomultiplier tube (EMI 9863) via an optical fiber bundle, which here only acted as a light pipe. The angle between the two detection arms could be varied from 10° to 160°, and the simultaneous dual-angle detection shortened the total measurement time by a factor of 2. The light source was a Spectra Physics 2020-03 argon ion laser operating at wavelength 488 nm and laser power 150 mW. For static light scattering, intensities were accumulated for 10 sec at scattering angles from 30° to 120° with 10° steps. For dynamic light scattering, correlation functions were determined from the fluctuating scattered intensity using a Brookhaven Instruments correlator (Model 2030). For calibration of the setup and to determine the absolute scattered intensity needed for static light scattering, the pure solvent benzene and polystyrene standards of known molar mass dissolved in cyclohexane have been used.

(iii) Methods used for data analysis:
To determine the particle characteristics by SLS, the refractive index increment dn_D/dc has to be known. This quantity was measured using a cylindrical light scattering cell deformed to permit refraction of the laser beam exiting the sample at the solution/air interface. The thus measured refraction led to the refractive index, using for calibration of the setup solvents of known refractive index at room temperature. Next, the refractive index of PETFE solutions in the solvent diisobutyl adipate was determined at a temperature of 240°C, with polymer concentrations ranging from 0 to 6e-3 g/mL, and the refractive index increment

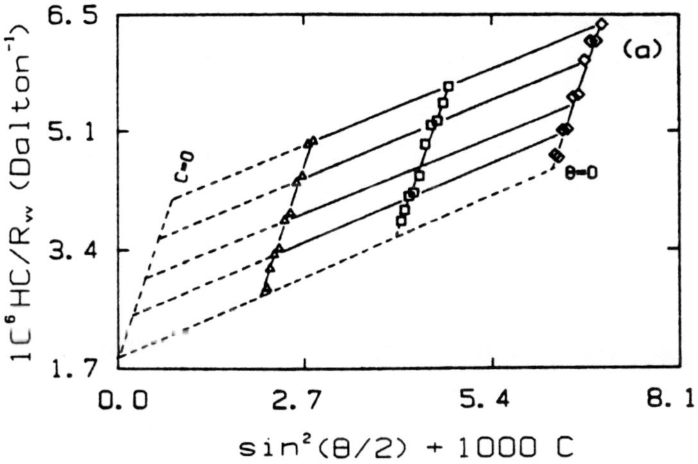

Fig. 5.23. Zimm plot of SLS data for PETFE in diisobutyl adipate at 240°C. Reprinted with permission from B. Chu and C. Wu, Macromolecules 20, 93–98, 1987, Copyright 1987 American Chemical Society (ref. [5.13]).

was determined from the slope of a linear curve plot of these data. Static light scattering results were analyzed using the Zimm equation (see Eq. 1.21) in Chap. 1.2).

(iv) Experimental results:
Figure 5.23 shows a Zimm plot for one of the PETFE-samples. Data evaluation using the intercept and the slopes of the two extrapolated ($q \rightarrow 0, c \rightarrow 0$) linear curves yields a weight-average molar mass $M_w = 5.4 \cdot 10^5\, g mol^{-1}$, a second Virial coefficient $A_2 = 1.97 \cdot 10^{-4}\, g^{-1} mol^{-1} mL^2$, and a z-average radius of gyration $\sqrt{\langle R_g^2 \rangle} = 45.4\, nm$. The three different sample concentrations studied were 2.09 mg/mL, 4.03 mg/mL and 6.25 mg/mL.

Using the SLS results from a series of samples with different molar masses, the authors determined the dependence of the radius of gyration on molar mass. As shown in Fig. 5.24, a scaling law was obtained with

$$\langle R_g^2 \rangle_z^{0.5} = 1.68 \cdot M_w^{0.60} \tag{5.35}$$

This relation suggests a coil topology of the PETFE chains in solution rather than a worm-like chain, as also seen from comparison with the fractal dimensions given in Table 1.1.

To complete their sample characterization, the authors also conducted DLS measurements. For data treatment, they used the Cumulant method (see Chap. 5.1, example A). Resulting apparent selfdiffusion coefficients were neither dependent on scattering angle nor on sample concentration in the studied regime, indicating a low sample polydispersity as well as the absence of particle interactions. The low polydispersity of the sample is also shown by the value of the

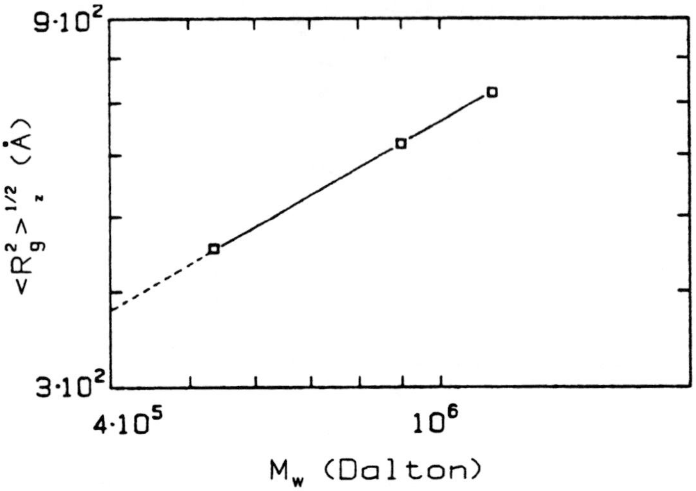

Fig. 5.24. Dependence of particle size on molar mass for PETFE in diisobutyl adipate at 240°C. Reprinted with permission from B. Chu and C. Wu, Macromolecules 20, 93–98, 1987, Copyright 1987 American Chemical Society (ref. [5.13]).

second Cumulant, which was smaller than 0.10. Combining the results from SLS and DLS, the authors have obtained a scaling law for the diffusion coefficient vs. molar mass, as shown in Fig. 5.25.

The power-law corresponding to the linear curve is given as:

$$\langle D_z \rangle = 3.35 \cdot 10^{-4} \cdot M_w^{-0.60} \tag{5.36}$$

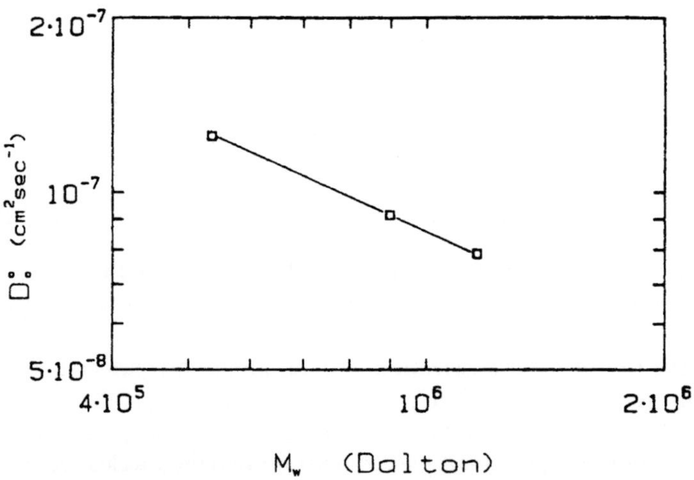

Fig. 5.25. Dependence of diffusion coefficient on molar mass for PETFE in diisobutyl adipate at 240°C. Reprinted with permission from B. Chu and C. Wu, Macromolecules 20, 93–98, 1987, Copyright 1987 American Chemical Society (ref. [5.13]).

if the diffusion coefficient is expressed in cm^2s^{-1} and the molar mass in Daltons. Interestingly, the exponent 0.60 is the same as that obtained for the SLS results, showing that the so-called ρ-ratio $\rho = R_g / R_H$ (see Chap. 1.3), and therefore the particle topology, corresponds to that of a random polymer coil in a good solvent (see Table 1.3) and does not depend on molar mass in the regime studied here. Further, since the diffusion coefficient did neither depend on scattering angle nor on particle concentration, in future one single dynamic light scattering measurement (the investigation of one sample at one scattering angle) should be sufficient to determine the molar mass of other PETFE polymers differing in molar mass from the samples studied here.

(v) Conclusions:
I have described here a practical example of how linear homopolymer chains in solution are characterized by static light scattering. Quantities determined range from molar mass and radius of gyration to sample topology and sample polydispersity, as well as sample-solvent interactions expressed by the second Virial coefficient A_2. For small scattering particles, where the fractal dimension of the scattering particles cannot be measured directly using a detailed analysis of the q-dependence of the particle form factor (as discussed in Chap. 1.2), the particle topology can be deduced by plotting the particle size, given by the particle radius of gyration or the hydrodynamic radius, vs. molar mass in a double logarithmic scale. If in this plot a linear curve is obtained, the slope should give the fractal dimension according to Eq. 1.40, and comparison with the values listed in Table 1.1 provides insight in the particle topology.

Example B (Ref. [5.14]):
Our second example illustrates how much more complicated the experimental characterization by static light scattering becomes if instead of a single homopolymer chain a block copolymer chain is investigated. These studies by Wu et al. provide an alternative to the conventional approach of light scattering characterization of heterogeneous segmented diblock copolymer chains: usually, characterization of heterogeneous copolymer chains affords static light scattering measurements of the sample in three different solvents which all differ in refractive index. This approach was first described by Bushuk and Benoit [5.15] and will also briefly be reviewed here.

(i) Samples and sample treatment:
The authors studied the segmented copolymer poly(ethyleneterephtalate-co-caprolactone) (PET-PCL), prepared in a two-step synthesis, that is esterification of terephtalic acid and ethyleneglycol followed by a poly-cocondensation of the ethyleneterephtalate with poly(caprolactone) of molar mass $M_n = 2{,}000$ g/Mol. Two samples with different average PET content were prepared. Here, we will focus on the sample with a lower content of 13% PET. This sample was precipitated from chloroform solution with methanol, leading to two fractions of similar composition but different average molar mass. For light scattering, the copolymer was

dissolved in two different analytical grade organic solvents: chloroform and THF, in a concentration range between 4×10^{-4} g/mL and 8×10^{-3} g/mL. All solutions were filtered through 0.22-μm Millipore filters to remove dust before the dynamic and static light scattering measurements.

(ii) Light scattering setup:
The light scattering measurements were conducted with a commercial setup (ALV DLS/SLS-5000 from ALV, Langen, Germany) equipped with an Ar^{+} laser (Coherent Innova 90) operated at wavelength 488 nm and laser power 400 mW. The incident laser light intensity was regulated with an attenuator (Newport M-925B) to suppress possible localized heating of the light scattering sample. Correlation functions measured with the ALV5000 multiple-τ digital correlator were accumulated until a net photo count of 10 e6: this means, for example, that at an average count rate of 200 kHz the accumulation time per correlation function would be 5 s. All light scattering measurements were done at 25°C.

(iii) Methods used for data analysis:
Characterization of single homopolymer chains by static light scattering is comparatively simple, using the Zimm plot for data analysis of the SLS results (see example 5.2.A). In contrast, segmented copolymer chains, especially in case of a heterogeneous composition varying with molar mass, provide a major challenge for SLS characterization. Bushuk and Benoit [5.15] have shown that at least three independent SLS experiments in different solvents are necessary to determine both molar mass distribution and composition profile of a polydisperse heterogeneous copolymer sample consisting of two monomer species A and B. For one given solvent, the Zimm-approach in this case only yields an apparent weight-average molar mass defined as:

$$M_{w,app} = \int_0^\infty f_w(M) \cdot \left(\frac{(dn_D/dc)(M)}{(dn_D/dc)_{app}} \right)^2 \cdot M dM \tag{5.37}$$

Equation 5.37 is based on the fact that the scattered intensity for a given particle depends both on its molar mass and its scattering contrast $(dn_D/dc)^2$. $f_w(M)$ is the weight distribution of the copolymer sample and $(dn_D/dc)(M)$ the refractive index increment for the sample fraction of molar mass M. The important point is that, as has been noted above, $(dn_D/dc)(M)$ may depend on the molar mass of the copolymer chain in case the sample consists of heterogeneous segmented diblock copolymer chains. $(dn_D/dc)_{app}$ is the apparent refractive index increment of the sample. Here, Wu et al. denoted $f_w(M)\cdot\left[(dn_D/dc)(M)/(dn_D/dc)_{app}\right]$ as the apparent weight distribution ($f_{w,app}(M)$). The respective refractive index increments from Eq. 5.37 depend on sample composition as:

$$(dn_D/dc)(M) = w_A(M)(dn_D/dc)_A + w_B(M)(dn_D/dc)_B \tag{5.38}$$

$$(dn_D/dc)_{app} = w_A \cdot (dn_D/dc)_A + w_B \cdot (dn_D/dc)_B \tag{5.39}$$

where $w_A(M)$ and $w_B(M)=1-w_A(M)$ are the weight fractions of monomer types A and B for a given copolymer chain of molar mass M and statistical weight $f_w(M)$. w_A and $w_B=1-w_A$ are the weight fractions for the whole copolymer sample usually known from synthesis of the copolymer, and $(dn_D/dc)_A$, $(dn_D/dc)_B$ are the refractive index increments for pure homopolymers of type A or B. Note that for a homogeneous copolymer the composition and therefore the refractive index increment is independent of the molar mass of the copolymer chain ($(dn_D/dc)(M)=(dn_D/dc)$). In this case we get:

$$M_{w,app}=\int_0^\infty f_w(M)\cdot MdM=M_w \tag{5.40}$$

On the other hand, for heterogeneous copolymer chains the refractive index increment may vary with molar mass. Following Wu et al. [5.14], the apparent molar mass determined from a Zimm analysis of SLS results for a given solvent-copolymer-pair can be expressed as:

$$M_{w,app}=M_w+2\cdot P\cdot\left(\frac{\Delta(dn_D/dc)}{(dn_D/dc)}\right)+Q\cdot\left(\frac{\Delta(dn_D/dc)}{(dn_D/dc)}\right)^2 \tag{5.41}$$

with

$$\Delta(dn_D/dc)=(dn_D/dc)_A-(dn_D/dc)_B, \tag{5.42}$$

$$M_w=\int_0^\infty f_w(M)\cdot MdM \tag{5.43}$$

$$P=\int_0^\infty f_w(M)\cdot M\cdot\left[w_A(M)-w_A\right]dM \tag{5.44}$$

$$Q=\int_0^\infty f_w(M)\cdot M\cdot\left[w_A(M)-w_A\right]^2 dM \tag{5.45}$$

The reader should note that (dn_D/dc) in Eq. 5.41 is an average refractive index increment determined experimentally for the given copolymer-solvent pair. Importantly, for a homogeneous copolymer the additional terms P and Q are equal to zero, since in this case $w_A(M)=w_A$. For a heterogeneous copolymer, on the other hand, we may obtain one pair of parameters $M_{w,app}$ and $\left[\Delta(dn_D/dc)\right]/(dn_D/dc)$ from SLS measurements and the determination of the respective refractive index increments $(dn_D/dc)_A$, $(dn_D/dc)_B$, using one given solvent. According to Eq. 5.41, we need at least three different results for $M_{w,app}$, obtained from SLS measurements of the heterogeneous copolymer chain in three different solvents, to determine the unknown quantities P and Q, and therefore also the third unknown quantity, the true weight average molar mass M_w. These three solvents must satisfy the following conditions: the average refractive increments (dn_D/dc) of the corresponding copolymer-solvent pairs should be as different as possible, the copolymer has to be soluble in all three solvents, and the final copolymer solutions should be as transparent as possible at the wavelength of the incident laser light. Finally, it has to be feasible to purify all copolymer solutions from dust. In practice, the choice of three suitable solvents becomes very difficult. The major problems rise from insufficient solubility of the copolymer, or too little light scattering contrast.

For this reason, Wu et al. developed a new procedure to characterize heterogeneous copolymers by light scattering, affording only two different solvents. Their idea was to replace the three sets of SLS experiments by a combination of SLS and DLS measurements affording only two different solvents, as described in more detail in the following: as has been shown above (see Eq. 1.57), DLS measurements of polydisperse samples at finite scattering angles yield an apparent diffusion coefficient. If we also include thermodynamic and hydrodynamic interactions of solvent and scattering particles, Eq. 1.57 has to be rewritten according to ref. [5.14] as:

$$D_{app}(q) = \Gamma/q^2 = D_z \cdot \left(1 + f \cdot R_g^{\,2} \cdot q^2\right) \cdot \left(1 + k_D \cdot c\right) \tag{5.46}$$

Here, c is the copolymer concentration, f is a dimensionless number and k_D, in analogy to static light scattering (see Eq. 1.21), is the so-called diffusion second Virial coefficient. The average scattered intensity, measured for the heterogeneous copolymer sample in the limit of infinite dilution and zero scattering angle ($c \to 0$, $q \to 0$) in a SLS experiment, according to Eq. 5.37 is given as:

$$\langle I \rangle \sim R = K \cdot c \cdot M_{w,app} = K \cdot c \cdot \int_0^\infty f_w(M) \cdot \left(\frac{(dn_D/dc)(M)}{(dn_D/dc)_{app}} \right)^2 \cdot M dM \tag{5.47}$$

On the other hand, the average scattered intensity also corresponds to the amplitude correlation function $g_1(\tau)$ measured in the short correlation time regime ($\tau \to 0$):

$$\langle I \rangle = \langle E(0) E^*(0) \rangle \sim g_1(0) = \int_0^\infty G(\Gamma) d\Gamma \tag{5.48}$$

with $G(\Gamma)$ the decay rate distribution of the polydisperse sample directly corresponding to a diffusion coefficient distribution $G(D)$. Combining Eqs. 5.47 and 5.48, Wu et al. derived the following relation:

$$\int_0^\infty G(D) dD \sim \int_0^\infty f_w(M) \cdot \left(\frac{(dn_D/dc)(M)}{(dn_D/dc)_{app}} \right)^2 \cdot M dM \tag{5.49}$$

For a copolymer species of given composition, the molar mass dependence of the diffusion coefficient can be expressed by a scaling law according to:

$$D = k_D \cdot M^{-\alpha_D} \tag{5.50}$$

with k_D, α_D two scaling constants independent of M for a given solvent and copolymer composition. Equation 5.49 can then be replaced by the following important expression:

$$f_{w,app}(M) = f_w(M) \cdot \left(\frac{(dn_D/dc)(M)}{(dn_D/dc)_{app}} \right)^2 \propto \frac{G(D)}{M} \cdot \frac{dD}{dM} \tag{5.51}$$

For a given set of scaling parameters k_D and α_D, according to Eq. 5.51 the diffusion coefficient distribution $G(D)$ can be converted into the apparent molar mass distribution $f_{w,app}(M)$. Repeating this procedure of analyzing the DLS data for two different solvents (and correspondingly two different sets of the scaling parameters k_D and α_D), one obtains two different apparent weight distributions $f_{w,app}(M)_1$ and $f_{w,app}(M)_2$. The ratio of these two apparent weight distributions according to Wu et al. is given as:

$$\frac{f_{w,app}(M)_1}{f_{w,app}(M)_2} = \left\{ \frac{(dn_D/dc)_2}{(dn_D/dc)_1} \cdot \frac{w_A(M)\cdot(dn_D/dc)_{A,1} + [1-w_A(M)]\cdot(dn_D/dc)_{B,1}}{w_A(M)\cdot(dn_D/dc)_{A,2} + [1-w_A(M)]\cdot(dn_D/dc)_{B,2}} \right\} \tag{5.52}$$

where the quantities $(dn_D/dc)_1$, $(dn_D/dc)_2$, $(dn_D/dc)_{A,1}$, $(dn_D/dc)_{A,2}$, $(dn_D/dc)_{B,1}$ and $(dn_D/dc)_{B,2}$ can be determined with a differential refractometer. Equation 5.52 allows to calculate the composition $w_A(M)$ and, correspondingly, $(dn/dc)(M)$, leading finally to $f_w(M)$ and M_w. The practice of this new procedure will be illustrated by briefly reviewing the experimental results given in ref. [5.14].

(iv) Experimental results:
Figure 5.26 shows the typical Zimm plot of the 13% PET-PCL sample obtained from SLS measurements for copolymer samples dissolved in chloroform in the concentration range between 7.9 e-4 g/mL and 3.95 e-3 g/mL.

Extrapolation of the light scattering data towards $q \to 0$ and $c \to 0$, as shown by the solid lines in Fig. 5.26, yielded the apparent sample characteristics summarized in Table 5.7.

Also shown in the table are the z-average diffusion coefficient $\overline{D}$ and the second Virial coefficient k_d as determined from an analysis of the DLS results according to Eq. 5.46. The decay rate distribution $G(\Gamma)$ has been determined by Laplace inversion of the DLS correlation function (see example 5.1 A). The average diffusion coefficients listed in Table 5.7 are corresponding to the average DLS decay rates according to

$$\overline{D} = \overline{\Gamma} \cdot q^2 = \int_0^\infty G(\Gamma) \cdot \Gamma d\Gamma \cdot q^2 \tag{5.53}.$$

Obviously, the solvent has a strong effect on the apparent sample characteristics.

To determine not only $M_{w,app}$ but M_w of the copolymer sample, Wu et al. had to measure the refractive index increments for the 13% PET-PCL copolymer

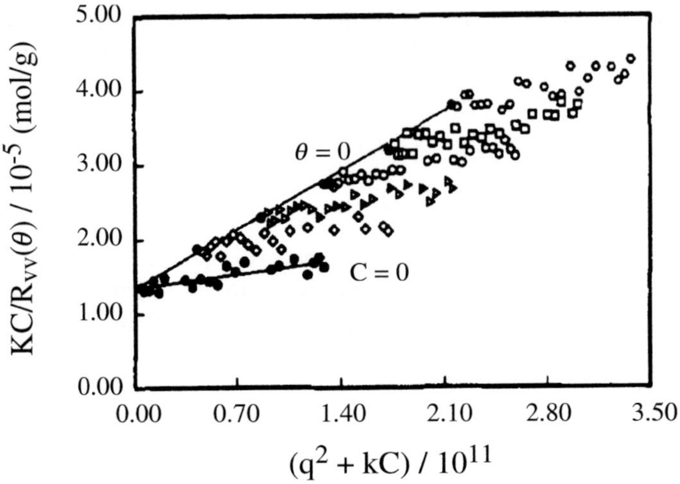

Fig. 5.26. Typical static Zimm plot of a 13% PET-PCL sample. Reprinted with permission from C. Wu, K.F. Woo, X. Luo and D.-Z. Ma, Macromolecules 27, 6055–6060, 1994, Copyright 1994 American Chemical Society (ref. [5.14]).

sample and PCL and PET homopolymers in the two different solvents chloroform and THF with a differential refractometer. The results (all in mL/g) were 0.071, 0.060, and 0.143 for the copolymer, the PCL homopolymer and the PET homopolymer in chloroform, and 0.090, 0.079, and 0.165 for the copolymer, the PCL homopolymer and the PET homopolymer in THF.

Regarding the data listed in Table 5.7, one may wonder why two different copolymer samples of, as assumed by Wu et al., similar composition but different molar mass were investigated both by SLS and DLS. As will be shown in the following, this was necessary to determine the still unknown scaling parameters k_D and α_D needed to convert the diffusion coefficient distribution, obtained by Laplace inversion of the DLS results, into the apparent molar mass distributions $f_{w,app}(M)_1$, $f_{w,app}(M)_2$. Figure 5.27 shows the diffusion coefficient distributions obtained for the two different fractions of the 13% PET-PCL copolymer in the solvent THF.

Table 5.7. Sample characteristics of 13% PET-PCL copolymer chains (low mass and high mass fraction) obtained from SLS and DLS. Reprinted with permission from C. Wu, K.F. Woo, X. Luo and D.-Z. Ma, Macromolecules 27, 6055–6060, 1994, Copyright 1994 American Chemical Society (ref. [5.14]).

Sample	Solvent	$10^{-4}\,M_{w,app}$ (g/mol)	$R_{g,app}$ (nm)	$10^7\,\overline{D}$ (cm^2/s)	k_d (mL/g)
13% PET-PCL	Chloroform	7.84	28	4.30	10
(low M)	THF	7.35	22	5.43	20
13% PET-PCL	Chloroform	27.9	30	2.33	90
(high M)	THF	29.6	28	3.11	80

According to Wu et al., the apparent molar mass, corresponding to the re sult from static light scattering, can be calculated from this diffusion coefficient distribution using the following relation:

$$M_{w,app} = \frac{k_D^{1/\alpha_D} \cdot \int_0^\infty G(D)dD}{\int_0^\infty G(D) \cdot D^{1/\alpha_D} dD} \tag{5.54}$$

Since the authors investigated two sample fractions of comparable composition, and therefore identical values for k_D and α_D, they derived Eq. 5.55 containing only one unknown quantity, α_D:

$$\frac{M_{w,app,1}}{M_{w,app,2}} = \frac{\int_0^\infty G_1(D)dD \cdot \int_0^\infty G_2(D) \cdot D^{1/\alpha_D} dD}{\int_0^\infty G_1(D) \cdot D^{1/\alpha_D} dD \cdot \int_0^\infty G_2(D)dD} \tag{5.55}$$

Combination of their static light scattering results $M_{w,app,1}$, $M_{w,app,2}$ given in Table 5.7 and their dynamic light scattering results $G_1(D)$, $G_2(D)$ shown in Fig. 5.27 allowed Wu et al. to determine a suitable value for α_D (= 0.59 for chloroform and 0.58 for THF).

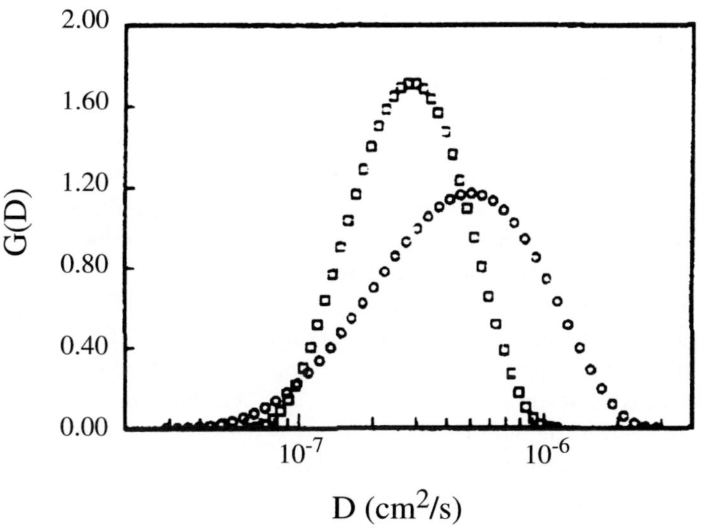

Fig. 5.27. Diffusion coefficient distributions of low-M (*circles*) and high-M (*squares*) 13% PET-PCL samples in THF. Reprinted with permission from C. Wu, K.F. Woo, X. Luo and D.-Z. Ma, Macromolecules 27, 6055–6060, 1994, Copyright 1994 American Chemical Society (ref. [5.14]).

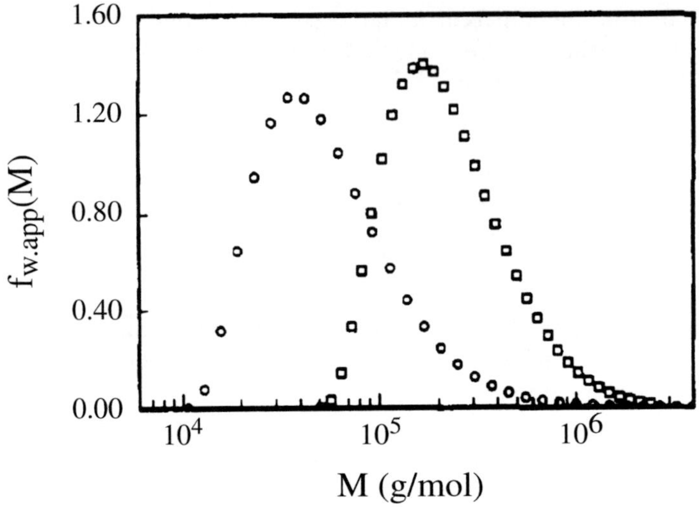

Fig. 5.28. Apparent weight distributions of low-M (*circles*) and high-M (*squares*) 13% PET-PCL samples in THF calculated from the diffusion coefficient distributions (**Fig. 5.27.**). Reprinted with permission from C. Wu, K.F. Woo, X. Luo and D.-Z. Ma, Macromolecules 27, 6055–6060, 1994, Copyright 1994 American Chemical Society (ref. [5.14]).

These values show that both THF and chloroform are equally good solvents for the copolymer chains. The theoretically predicted value of α_D for flexible linear polymer chains in a good solvent is 0.60. Once α_D was determined, it can be used to calculate the second unknown (k_D), using Eq. 5.50. The corresponding results were $k_D = 5.3e-4$ for chloroform and $k_D = 6.0e-4$ for THF. With these parameters, the apparent molar mass distributions $f_{w,app}(M)$ could be determined from the respective distributions $G(D)$ using Eqs. 5.50 and 5.51. These results from ref. [5.14] are shown in Fig. 5.28.

Using the ratio of apparent molar mass distributions $f_{w,app}(M)$, measured in the two different solvents THF and chloroform, and the refractive index increments given above, the authors were now ready to calculate the copolymer chain composition distribution $w_A(M)$ according to Eq. 5.52. Their respective results for the two different sample fractions are shown in Fig. 5.29.

Obviously, the lower mass copolymer fraction contains fewer PET segments than the higher molar mass fraction. Here, one might criticize that the authors had assumed a similar sample composition for determination of their values k_D and α_D, which obviously does not seem to be the case. We may conclude that the approach is only valid if the parameters k_D and α_D do not depend on sample composition in the regime $w_A(M)$ displayed in Fig. 5.29, that is $5wt\% < w_{PET}(M) < 15wt\%$. From $w_{PET}(M)$, next Wu et al. could calculate the refractive index increment in dependence of molar mass $(dn_D/dc)(M)$, and finally the true molar mass distribution $f_w(M)$. Using this mass distribution, $M_w = 6.7e4$ g/Mol and a sample polydispersity $M_w/M_n = 1.9$ had been calculated for the copolymer. For comparison, Wu et al. also applied the standard

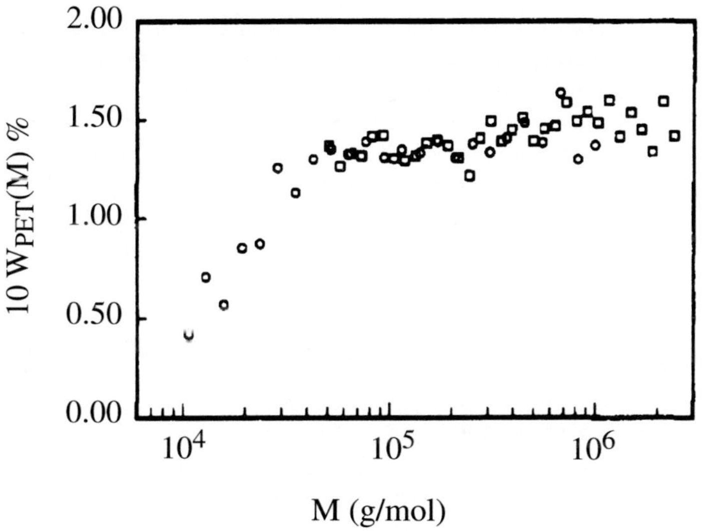

Fig. 5.29. Calculated composition distributions of low-M (*circles*) and high-M (*squares*) 13% PET-PCL samples. Reprinted with permission from C. Wu, K F. Woo, X. Luo and D.-Z. Ma, Macromolecules 27, 6055–6060, 1994, Copyright 1994 American Chemical Society (ref. [5.14]).

three-solvent method suggested by Bushuk and Benoit. Fitting their apparent molar masses determined by SLS, as described, to Eq. 5.41, they obtained $M_w = 6.6e4$ g/Mol, which is in excellent agreement with the results obtained from their alternative new approach affording only light scattering in two different solvents.

(v) Conclusions:
As Wu et al. already stated in their article, the experimental errors and uncertainties associated especially with the inverse Laplace transformation of the DLS data are rather large. Also, it is not clear (from ref. [5.14]) if the authors determined their diffusion coefficient distributions by extrapolation to scattering angle zero, as should be done (see above). For these reasons, the calculated sample composition $w_{PET}(M)$ can only be an estimated value. Nevertheless, the copolymer weight-average molar mass determined by the new approach agreed very well with the results obtained from the standard three-solvent SLS procedure [5.15]. Finally, one should note that the apparent molar masses determined from SLS in chloroform or THF both differed strongly from the true molar mass (see Table 5.7). This underlines the necessity of multiple solvent measurements for the characterization of heterogeneous copolymers by laser light scattering, using either the standard three-solvent SLS approach [5.15] or the new procedure presented here [5.14].

Example C (Ref. [5.16]):
Here, we will briefly review static light scattering studies concerning the formation of diblock copolymer micelles from single copolymer chains in dilute solution.

An important quantity describing this phenomenon is the critical micelle concentration (cmc). At copolymer concentrations lower than the cmc, the sample exists in the form of single copolymer chains. With copolymer concentration increasing above the cmc, spherical micelles containing several dozen to several hundred single copolymer chains are formed, whereas the concentration of the single copolymer chains is kept constantly at the cmc value. This dramatic change in sample topology can easily be observed in a static light scattering experiment in the regime of zero scattering angle, since the scattered intensity detected in this limit depends both on number density and squared mass of the scattering particles.

(i) Samples and sample treatment:
Three different types of symmetric polystyrene(PS)-polyisoprene(PI) block copolymers, all of total molar mass 24,000 g/Mol and prepared by living anionic polymerization, were studied by Schädler et al.: (1) a standard noncharged PSPI-copolymer (denoted H24 in the following), (2) a copolymer endcapped with a $SO_3^-Li^+$-group at the PI chain end (sample S24), and (3) a so-called α,ω-macrozwitterionic copolymer with a $SO_3^-Li^+$-group at the PI chain end and a $N(CH_3)_3^+Br^-$-group at the PS chain end (Z24). All three copolymers were dissolved in HPLC grade dimethylacetamide (DMAc) and allowed to stand for at least 12 h before the light scattering measurements to ensure equilibrium. For purification of the sample solutions, 0.45 μm PTFE/PP syringe filters (Schleicher & Schuell) were used.

(ii) Light scattering setup:
SLS measurements were performed with a home-built goniometer-based light scattering setup, which allows one to detect the time-averaged scattered intensity in an angular range of 10°–140°, using an EMI-type photomultiplier as detection unit. An Ar^+ laser operating at 488 nm was used as the light source. Refractive index increments needed for SLS measurements were determined separately with a home-built, high-precision scanning Michelson interferometer [5.17]. The refractive index increment of all three copolymer samples was identical (0.120 mL/g), the refractive indices themselves are (all at temperature 293 K) $n_{D,PS} = 1.583$, $n_{D,PI} = 1.521$ and $n_{D,DMAc} = 1.438$.

(iii) Methods used for data analysis:
As discussed in detail in the preceding example B, static light scattering from copolymer solutions only yields an apparent molar mass. As correctly claimed by Schädler et al., the difference between $M_{w,app}$ and M_w here should be comparatively small since the refractive indices both of the PS and the PI block are much larger than that of the solvent. Also, the polydispersity and composition heterogeneity of copolymers prepared by living anionic polymerization are rather small. The authors therefore treated their SLS data in the usual way like results obtained from homopolymer solutions. The major difference is that their copolymer at higher concentration forms well-defined micellar aggregates. As a consequence, the total scattered intensity or Rayleigh ratio in this transition

regime is a combination of scattering from single copolymer chains (unimers) and scattering from the micelles:

$$R = R_u + R_m \tag{5.56}$$

Using the Zimm equation (Eq. 1.21), the total scattered intensity at zero scattering angle then can be expressed as:

$$R(q=0) = \frac{K \cdot c_m \cdot (M_{w,app})_m}{1 + 2 \cdot A_2 \cdot c_m \cdot (M_{w,app})_m} + \frac{K \cdot c_u \cdot (M_{w,app})_u}{1 + 2 \cdot A_2 \cdot c_u \cdot (M_{w,app})_u} \tag{5.57}$$

where the index m corresponds to micelles and the index u corresponds to unimers. Usually, the molar mass of the micelles is much larger than that of the unimers (typical aggregation numbers of copolymer micelles are about 100), and the second Virial coefficient A_2 and the unimer concentration c_u are comparatively small. Equation 5.57 for these reasons can be rewritten as:

$$R(q=0) = \frac{K \cdot c_m \cdot (M_{w,app})_m}{1 + 2 \cdot A_2 \cdot c_m \cdot (M_{w,app})_m} + K \cdot c_u \cdot (M_{w,app})_u \tag{5.58}$$

On the other hand, the total copolymer concentration is given as $c = c_m + c_u$, with $c_u = cmc$ (see above). Therefore, the reduced light scattering intensity can be expressed as:

$$\frac{K \cdot c}{R} = \frac{c}{(M_{w,app})_u + \frac{(M_{w,app})_m \cdot (c - cmc)}{1 + 2 \cdot A_2 \cdot (M_{w,app})_m \cdot (c - cmc)}} \tag{5.59}$$

Equation 5.59 is the fitting equation to analyze the SLS data presented in ref. [5.16], allowing to determine the cmc, A_2, $M_{w,app} \approx M_w$, and finally the average aggregation number of the micelles $N_{agg} = M_w / M_u$.

(iv) Experimental results:

Figure 5.30 shows the reduced scattered intensity extrapolated to zero scattering angle for the three different copolymer samples, plotted versus copolymer concentration.

In the dilute regime slightly above the cmc, Kc/R is strongly decreasing with increasing micellar concentration, reflecting the increase in average molar mass. Well above the cmc, which is in case of our example at copolymer concentrations c >1 g/L, the average molar mass remains constant. In this regime, the contribution of the unimers to the overall scattered intensity becomes negligible, and Kc/R reflects the inverse average molar mass of the micelles only. Fitting the data displayed in Fig. 5.30 to Eq. 5.59, Schädler et al. obtained the results summarized in Table 5.8.

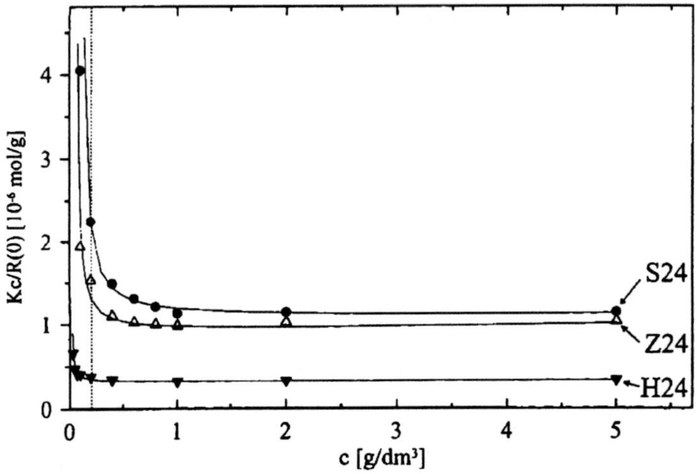

Fig. 5.30. Reduced light scattering intensity Kc/R extrapolated to zero scattering angle ($= M_w^{-1}$) versus copolymer concentration for different types of copolymers in DMAc. Reprinted with permission from V. Schädler, C. Nardin, U. Wiesner and E. Mendes, J.Phys.Chem.B 104, 5049–5052, 2000, Copyright 2000 American Chemical Society (ref. [5.16]).

Table 5.8. Sample characteristics of PSPI copolymer micelles obtained from SLS. Reprinted with permission from V. Schädler, C. Nardin, U. Wiesner and E. Mendes, J.Phys.Chem.B 104, 5049–5052, 2000, Copyright 2000 American Chemical Society (ref. [5.16]).

Sample	cmc (g/dm^3)	A_2 (10^{-9} dm^3mol/g^2)	$M_{w,app}$ (10^6 g/mol)	N_{agg}
H24	0.0177	1.62	3.17	132
S24	0.110	7.01	0.950	39.6
Z24	0.065	5.44	1.12	46.5

Obviously, the charged copolymer chains show a much higher cmc and lower average aggregation number than their noncharged counterpart sample H24. This is corresponding to the larger A_2 values of the charged copolymer chains, indicating their higher solubility in the solvent DMAc compared to the non-charged sample H24. A more detailed discussion of these results is beyond the scope of this book, and the interested reader is referred to ref. [5.16].

(v) Conclusions:

Example C illustrates how static light scattering is used to characterize copolymer micelles, a very important architecture in modern nanoparticle research. It should be noted that all results presented here are only apparent values for the reasons discussed in example B. However, "true" and apparent values should be comparable in the present case of rather well-defined copolymer chains prepared by living anionic polymerization.

Example D (Ref. [5.18]):
This example introduces light scattering characterization of a more complicated sample architecture: branched polymers in solution. As in the previous examples, sample characterization as given by M, R_g, A_2, and R_H will be one issue. More importantly, Galinsky and Burchard tried to explore the branching topology itself from their scattering results, using a detailed analysis of the q-dependence of the particle form factor. Their procedure will be briefly reviewed here.

(i) Samples and sample treatment:
Several starch fractions were prepared by a controlled acid degradation of potato starch granules. The acid causes selective cleavage of the glycoside bonds in the amorphous interlamellar sheets of the granules, resulting in fractions which essentially retain all features of native amylopectin. By this method also called "linterization", seven branched samples with suitable size range for detailed form factor analysis by static light scattering were obtained. These samples were dissolved in 0.5 N NaOH at concentrations ranging from a very dilute system to 210 g/L. For light scattering, the solutions were filtered three times through Millipore filters of 1.2, 0.8, 0.45, and 0.2-µm pore size depending on molar mass of the scattering particles, and the third filtration was conducted directly into cylindrical light scattering cuvettes of inner diameter 0.8 cm. These cells had been cleaned using the standard "distilled acetone technique" as described in previous examples. The refractive index increment dn_D/dc of these solutions, needed to obtain quantitative SLS results, was measured by refractometry as 0.142.

(ii) Light scattering setup:
The light scattering setup was a modified and fully computerized SOFICA photogoniometer (Baur Instrumentenbau, Germany), equipped either with a 2-mW HeNe laser with wavelength $\lambda = 632$ nm, or a 5-mW Ar^+ laser ($\lambda = 488$ nm), both from Uniphase. The scattered light intensity was accumulated in an angular range from 30° to 145° in steps of 5°.

(iii) Methods used for data analysis:
Molecular parameters of the samples were determined from static light scattering in very dilute solutions using the Zimm equation. The more detailed analysis of the particle form factor, allowing one to extract information on the branching structure of the particles, will be shown in the next section where experimental results are presented.

(iv) Experimental results:
Scattering data obtained from very dilute solutions have first been analyzed using the Zimm equation. The resulting sample parameters are summarized in Table 5.9.

The dependence of absolute scattered intensity of these samples on scattering vector is shown on a double-logarithmic scale in Fig. 5.31.

Also presented in the data plot are the normalized data obtained by (1) dividing the scattered intensities $I(q)$ by the scattered intensity determined by extrapolation to $q \rightarrow 0$, and (2) rescaling the scattering vector with the average

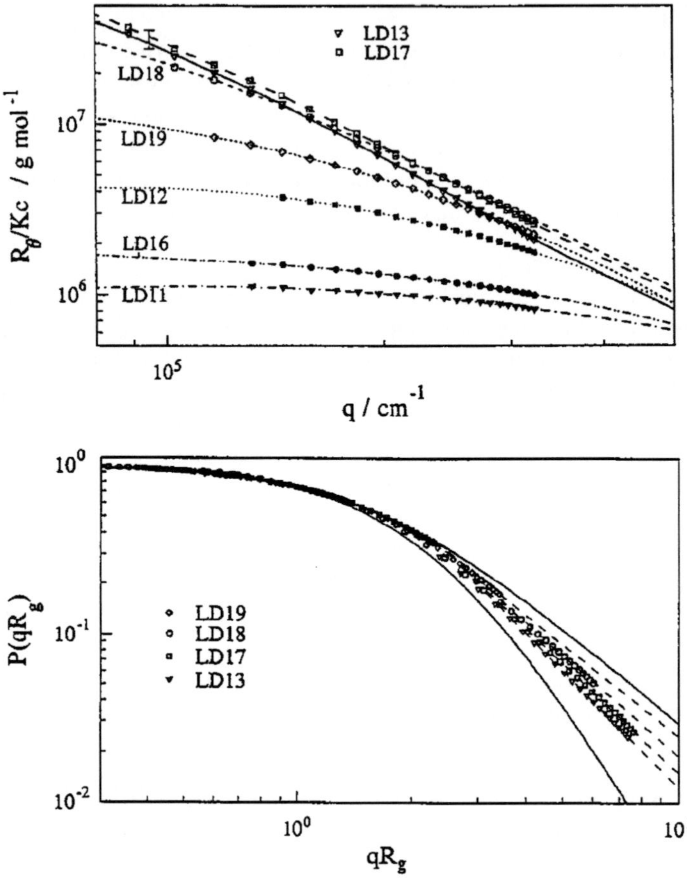

Fig. 5.31. *q*-dependence of absolute (*top*) and normalized (*bottom*) scattered intensities for degraded starch samples. Reprinted with permission from G. Galinsky and W. Burchard, Macromolecules 30, 4445–4453, 1997, Copyright 1997 American Chemical Society (ref. [5.18]).

particle size as given by the radius of gyration. In case of the very dilute solutions studied here, where particle interactions leading to structure factor contributions in the scattered intensity are negligible, the normalized scattered intensities shown at the bottom of Fig. 5.31 correspond to the particle form factor $P(q)$. In case the structures are self-similar, that is the particle topology does not vary with molar mass of the scattering particles, the normalized particle form factor should be independent of sample molar mass and therefore universal. This universality is indeed given for the starch samples studied here up to values of $qR_g = 2$, whereas at larger q (corresponding to smaller length scales or a more detailed probing of the sample structure) systematic deviations occur. Nevertheless, all curves seem to show a linear behavior corresponding to the power law expected for light scattering from fractal or self-similar objects at

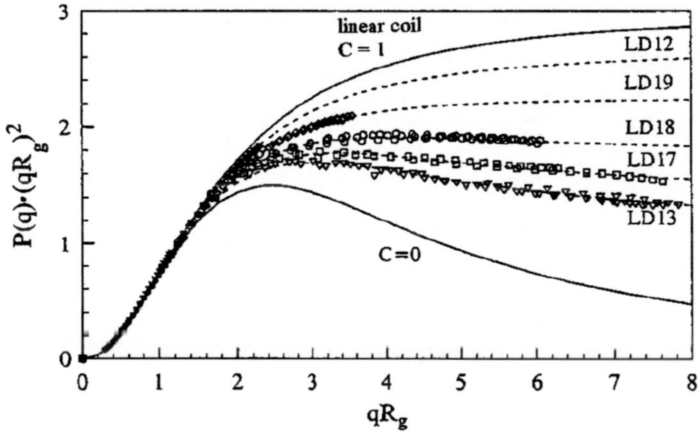

Fig. 5.32. Kratky plot of the scattering data shown in **Fig. 5.31**. Reprinted with permission from G. Galinsky and W. Burchard, Macromolecules 30, 4445–4453, 1997, Copyright 1997 American Chemical Society (ref. [5.18]).

Table 5.9. Particle characteristics of branched starch samples as determined from Zimm analysis. Reprinted with permission from G. Galinsky and W. Burchard, Macromolecules 30, 4445–4453, 1997, Copyright 1997 American Chemical Society (ref. [5.18]).

Sample	$10^{-6}\,M_w$ (g/mol)	R_g (nm)	$10^4\,A_2$ [(mol cm^3)/g^2]
LD11	0.92	36	1.00
LD16	1.87	48	0.60
LD12	5.20	70	0.28
LD19	14.5	113	0.13
LD18	43	180	0.082
LD17	64	190	0.060
LD13	97	233	0.025

larger q. The slopes of the curves shown in Fig. 5.31, corresponding to the fractal dimensions according to Eq. 1.41, vary between 1.85 and 2.32.

The minor deviations in the particle form factors of the different sample fractions at higher q-values are better visualized in the so-called Kratky plot shown in Fig. 5.32.

This plot emphasizes the large-q regime by multiplying the scattered intensity with $\left(qR_g\right)^2$. As also shown in Fig. 5.32, the form factors could be well described by a simple equation using only one fit parameter C:

$$P(q)=\frac{1+\left(C/3\right)\left(qR_g\right)^2}{\left[1+\left(\left(1+C\right)/6\right)\left(qR_g\right)^2\right]^2} \tag{5.60}$$

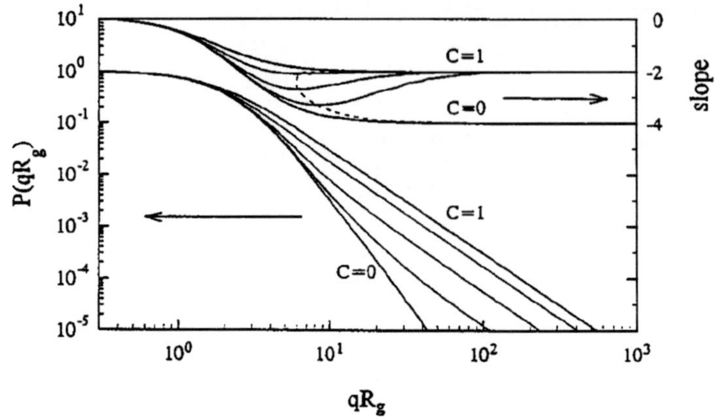

Fig. 5.33. Double logarithmic plot of particle form factors predicted for branched macromolecules (*lower part*), and the slopes of these curves (*upper part*). Reprinted with permission from G. Galinsky and W. Burchard, Macromolecules 30, 4445–4453, 1997, Copyright 1997 American Chemical Society (ref. [5.18]).

This equation has been derived originally to fit scattering data from samples prepared by polycondensation of trifunctional monomers, neglecting the influence of excluded volume. The parameter C is related to the branching probability and increases with molar mass, showing that the starch samples obviously differ in branching architecture from each other.

It remains an open question if the starch samples are, although not self-similar, fractal objects, meaning each fraction itself has a self-similar internal structure. To address this problem, particle form factors, which correspond to the fitting functions shown in Fig. 5.32, have been calculated. The extreme cases $C = 0$ (corresponding to a linear polymer coil) and $C = 1$ have been plotted in a double logarithmic plot of $P(q)$ versus qR_g, together with the q-dependence of the slope of these curves (see Fig. 5.33).

For the calculated form factors, a minimum slope is reached at qR_g equalling about 10, as indicated by the dashed curve in Fig. 5.33. This scattering vector regime corresponds to the maximum q range covered by the experimental data (see Fig. 5.31). However, at higher q values, only accessible either in simulations as shown here or by switching to a different scattering technique using radiation of smaller wavelength than light (X-ray or neutron scattering), the slope approaches a value of −2.0 which is characteristic for a linear polymer chain in the unperturbed state. This q range corresponds to a very small length scale where indeed only linear chain sections are visible, a large group of which have to be nonbranched outer chains of the hyperbranched starch particles. For this reason, only the limited q range of the light scattering experiment may have caused the impression that the branched starch samples are fractal objects. Here, the concept of fractal dimensions holds only in an intermediate narrow q-regime covering less than one decade around $qR_g = 10$, and therefore only encompassing several branching generations.

Although particle characterization by light scattering usually is limited to the regime of very dilute samples for the reasons frequently mentioned above, and formalisms like the Zimm equation are expected to fail at higher sample concentrations, Galinsky and Burchard have nevertheless tried to apply the Zimm approach to the semidilute concentration regime. These interesting results will be reviewed in the following. For nondilute solutions, the intercept of the absolute scattered intensity at scattering angle $\theta = 0$ compared to the simple Zimm equation (Eq. 1.21) has to be expanded, using additional higher order Virial coefficients $A_3, \ldots$ according to:

$$\left(\frac{Kc}{R_{\theta=0}}\right)_c = \frac{1}{M_w}\left[1 + 2A_2 M_w c + 3A_3 M_w c^2 \ldots\right] \equiv \frac{1}{M_{app}(c)} \tag{5.61}$$

The forward scattered intensity in this case yields only an apparent molar mass, which depends strongly on sample concentration. In addition, the initial slope of the inverse scattered intensity plotted versus scattering vector q^2 always allows one to determine a radius of gyration R_g or, more general, a distance length ξ according to:

$$R_{g\ app}^2(c) \equiv 3\frac{initial\,slope}{intercept} \equiv 3\xi^2(c) \tag{5.62}$$

To describe the influence of intermolecular interactions, the authors used two different approaches for concentrations up to the particle overlap concentration c^*, and for samples with much larger concentrations:

(a) For $c < c^*$, besides A_2 no higher order Virial coefficients are needed according to Galinsky and Burchard, and the Zimm equation, together with the apparent quantities defined above (Eqs. 5.61, 5.62), yields the following expression:

$$R_{g\ app}^2(c) \cong R_g^2 \frac{M_{app}(c)}{M_w} = \frac{R_g^2}{\left[1 + 2A_2 M_w c\right]} \tag{5.63}$$

with $R_{g\ app}^2(c)/M_{app}(c)$ the initial slope of the inverse absolute scattered intensity plotted vs. q^2 for a sample of given concentration *c*.

(b) For higher concentrations $c > c^*$ in the so-called semidilute regime, the validity of the Zimm expression, even if expanded with higher order Virial coefficients, is no longer given. In this regime of overlapping polymer coils, the correlation length should not any longer represent the size of individual scattering particles but rather an average "mesh size" of a continuous polymer network constituted of entangled polymer chains. Therefore, ξ should become independent of the molar mass of the scattering particles and should only depend on sample concentration. According to the scaling law found for linear polymer

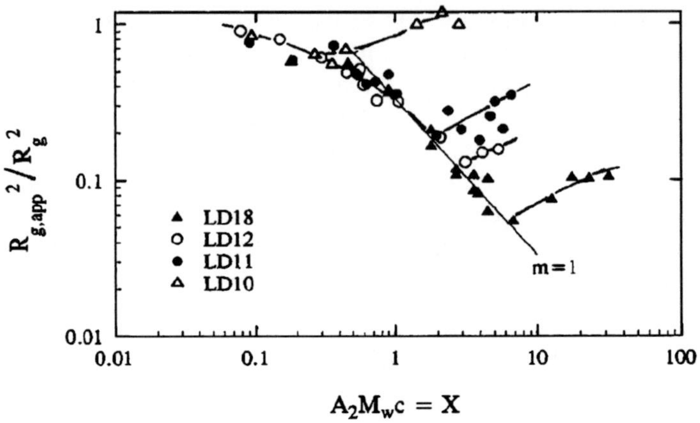

Fig. 5.34. Concentration dependence of the normalized apparent radius of gyration for different starch sample fractions. Reprinted with permission from G. Galinsky and W. Burchard, Macromolecules 30, 4445–4453, 1997, Copyright 1997 American Chemical Society (ref. [5.18]).

chains in the semidilute regime, a similar dependence has been assumed for the branched system investigated here:

$$R_{g,app}(c) = R_g\left(c/c^*\right)^m \tag{5.64}$$

with the exponent m being yet undetermined. As not shown here, SLS from very dilute solution has yielded $R_g \sim M^{0.39}$. On the other hand, the overlap concentration depends on the second Virial coefficient as $c^* = (A_2M_w)^{-1}$. Including the molar mass dependence of the second Virial coefficient, a value $m = -1.0$ has been calculated. This corresponds to an asymptotic decrease of $R_{g,app}(c)$ with c^{-1}, compared to $c^{-0.75}$ found for linear chains. Figure 5.34 shows the apparent radius of gyration determined from initial slope and intercept of the inverse scattered intensity, normalized with the radius of gyration measured by SLS from very dilute sample solutions and plotted vs. renormalized sample concentration $X = A_2 \cdot M_w \cdot c$.

As can be seen, for $c > c^*$, corresponding to $A_2M_wc > 1$ and $c^* = 2.8\,g/L$, the apparent radius of gyration universally scales with c^{-1} as predicted. The unexpected sharp increase in $R_{g,app}$ at higher concentrations here is attributed to the formation of clusters by particle association.

Next, Galinsky and Burchard introduced a dimensionless apparent particle form factor according to:

$$P_{app}(q,c) \equiv \frac{S(q,c)}{S(q=0,c)} \tag{5.65}$$

In Eq. 5.65, $S(q,c)$ is the scattering intensity detected at finite scattering angle, and $S(q=0,c)$ is the forward scattering. Surprisingly, when plotting this quantity $P_{app}(q,c)$ versus $qR_{g,app}(c)$, all data measured for a given sample at six

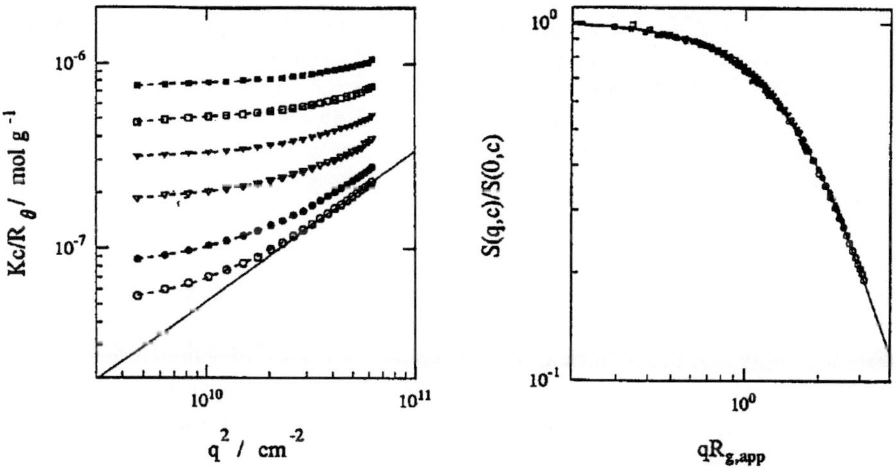

Fig. 5.35. Plot of the reciprocal scattered intensity measured for starch sample LD18 at six concentrations up to $c \approx 5c^*$ (*left*), and normalized form factor extracted from these data (*right*). Reprinted with permission from G. Galinsky and W. Burchard, Macromolecules 30, 4445–4453, 1997, Copyright 1997 American Chemical Society (ref. [5.18]).

different concentrations, ranging from 1.3 mg/mL to 12.7 mg/mL, which is well above $c^* = 2.8 mg/mL$, superimpose onto a single master curve. This master plot importantly corresponds to the particle form factor measured for the very dilute sample, as shown in Fig. 5.35.

On the other hand, the individual absolute scattered intensities, plotted as in verse scattered intensities vs. q^2 according to the Zimm equation, show a strong effect of sample concentration on both absolute values and curvature. One may conclude that the shape of the macromolecules themselves is not changed by particle interactions in the present case. As a consequence, once the two limiting scattering curves at $c = 0$ and $q = 0$ are known, the whole data set of scattered intensities detected for finite sample concentrations and finite scattering vectors can be reconstructed from the parameters R_g, M_w, A_2, c^*, m, where the later two are needed only in case concentrations well above the overlap concentration are concerned. Also, intercept and initial slope can be determined for a SLS data set obtained at any concentration, yielding $M_{app}, R_{g,app}$ and $S(q,0)$. This information is sufficient to construct the particle form factor usually only measured for very dilute particle solutions. To determine the "true" particle molar mass, the second Virial coefficient is needed and therefore light scattering measurements at various sample concentrations, but only in the small q-regime, are necessary.

(v) Conclusions:

Example D has not only provided a more detailed analysis of the q-dependence of the particle form factor compared to the simple Zimm equation, which is valid in the asymptotic regime of small q and small c. Also, it has demonstrated that particle interaction contributions to the measured scattered intensity can be eliminated in case the scattering particle itself is not changing in size and shape with increasing sample concentration. This data correction may yield the single particle form factor even for measurements of samples in the semidilute concentration regime by just determining apparent radius of gyration and apparent molar mass from initial slope and intercept of the data, plotted as inverse scattered intensity versus q^2, respectively, as shown in Fig. 5.35.

Example E (Ref. [5.19]):

Our next example will deal with light scattering experiments on vesicular structures in solution. The main issue of the authors was to separate effects of polydispersity and nonspherical particle shape on the angular dependence of the static scattered intensity by using a combination of static and dynamic light scattering. As will be shown, reliable particle characterization had become possible by this procedure.

(i) Samples and sample treatment:

Two different types of unilamellar vesicles were prepared from the commercially available lipid molecules 1,2-Dioleoyl-*sn*-glycero-3-phosphocholine (DOPC) and 1-stearoyl-2-oleoyl-*sn*-glycero-3-phosphocholine (SOPC) by extrusion: first, a lipid film was prepared by casting from $MeCl_2$-solution within a 100-mL, round-necked flask. This film was then dispersed in a morpholinopropane sulfonic acid (MOPS) buffer solution of pH 7.4 (20 mm MOPS, 100 mm NaCl, + NaOH to adjust the pH) at concentration 50 mg/mL. The dispersion was repeatedly freeze-thawed by alternating immersion in liquid nitrogen and warm water to transform originally multilamellar vesicles into unilamellar vesicles. Next, the vesicle dispersion was extruded with a hand-held extrusion device in two steps: first seven times through a 200-nm pore polycarbonate filter, and then multiple times through a 100-nm pore filter. Light scattering measurements were performed within hours of extrusion after the vesicle dispersions were diluted to a final concentration of approximately 0.05 mg/mL. Since no details of this dilution process are given in ref. [5.19], we assume that additional buffer solution has simply been filtered into the light scattering cuvette.

(ii) Light scattering setup:

The setup used for DLS consisted of a diode-pumped frequency doubled Nd:YAG laser (Coherent 532 DPSS) with wavelength 532 nm and a Brookhaven digital correlator with software package for data accumulation and analysis (BI-9000 AT and 9KDLSW control program). The intensity correlation functions $g_2(\tau)$ were measured at scattering angle 90° with a linear spacing of correlation times, and analyzed with a discrete Laplace inversion routine to yield the intensity-weighted size distribution of the vesicles. This size distribution was converted into a vesicle number distribution as described in the next section. The SLS setup consisted of a Malvern 4700 spectrometer equipped with a Siemens 632.8 nm HeNe laser.

(iii) Methods used for data analysis:
The particle form factor of monodisperse vesicles is given as:

$$P(q)=\left[\frac{3}{R_o^3-R_i^3}\right]^2\left[R_o^3\frac{j_1(qR_o)}{qR_o}-R_i^3\frac{j_1(qR_i)}{qR_i}\right]^2 \tag{5.66}$$

Assuming an average vesicle radius R and a vesicle shell thickness t, the outer and inner radius of the vesicle are given as $R_o=R+t/2$ and $R_i=R-t/2$. $j_1(x)$ is the first-order spherical Bessel function defined as:

$$j_1(x)=\frac{\sin x}{x^2}-\frac{\cos x}{x} \tag{5.67}$$

For small scattering angles or small values of qR, the form factor can be expressed according to the so-called Guinier approximation as

$$P(q)\approx\exp\left(-q^2R_g^2/3\right) \tag{5.68}$$

The typical q range of light scattering experiments is 0.002 nm^{-1} to 0.03 nm^{-1}, whereas vesicles prepared by extrusion have average radii ranging from 20 to 100 nm, corresponding to a qR range of 0.04 to 0.6 for 20 nm vesicles, and to a qR range of 0.2 to 3 for 100 nm vesicles. This means that the first minimum of the particle form factor will not be visible in a standard static light scattering measurement, as also seen from the form factors calculated within the experimental q range for different vesicle sizes shown in Fig. 5.36.

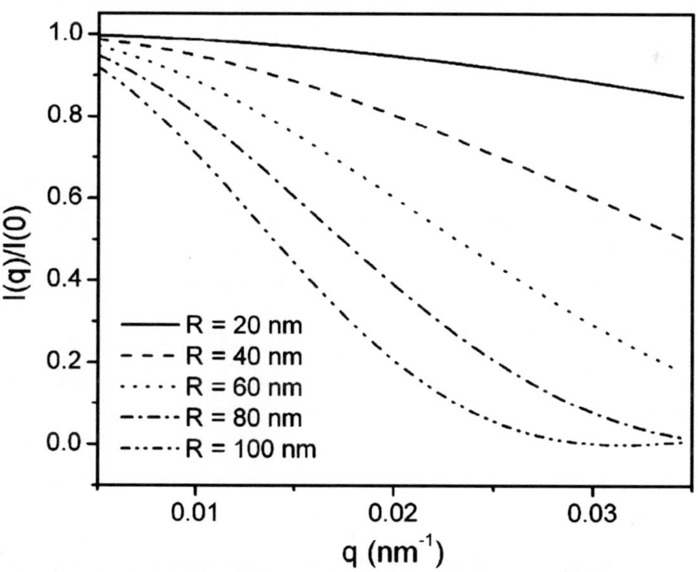

Fig. 5.36. Simulated particle form factors for monodisperse vesicles of various sizes and shell thickness 4 nm. Reprinted with permission from J. Pencer and F.R. Hallett, Langmuir 19, 7488–7497, 2003, Copyright 2003 American Chemical Society (ref. [5.19]).

The vesicle size can be extracted from the measured particle form factors using a couple of different approaches: (1) the full form factor including the Bessel function (Eq. 5.66), in the following called RGDF; (2) the thin-shell approximation of the form factor corresponding to the expression for ideal hollow spheres (shell thickness $t=0$), abbreviated as RGDT and given as:

$$P(q) \approx \left[\frac{\sin(qR)}{qR}\right]^2 \tag{5.69}$$

and (3) several forms of Guinier plots: $-\ln\left(I(q)\right)$ vs. q^2 (G1), $I(0)/I(q)$ versus q^2 (G2) and $\left[I(0)/I(q)\right]^{0.5}$ vs. q^2 (G3). Whereas the form factor analysis methods RGDF and RGDT yield the vesicle size *R*, the slopes of the Guinier plots according to Eq. 5.68 give the radius of gyration, which for spherical vesicles is given as:

$$R_g^{\,2} = \frac{3}{5} R_o^{\,2} \frac{1-\left(R_i/R_o\right)^5}{1-\left(R_i/R_o\right)^3} \tag{5.70}$$

Here, one should note that for vesicles with an infinitely thin shell $t=0$ one obtains $R_g = R$. From Guinier fits of their simulated data, Pencer and Hallett found the best agreement between fit results and theoretically expected vesicle size for method G1, whereas methods G2 and G3 lead to a considerable overestimation of the particle size. Also, the deviation increases with particle size, which is to be expected since the Guinier approximation should be strictly valid only for small values of *qR*. Limiting the fitting range to $qR<1$, the Guinier fits gave more reasonable results. This, however, causes a problem in experimental practice, where the vesicle size and therefore *q* range suitable for Guinier analysis are not known a priori. The solution is to vary the *q* range of the fit from $q=0$ to $q_{\max}$ by decreasing $q_{\max}$ until the value of R_g obtained from data fitting remains constant. Importantly, a detailed form factor fit using RGDF and RGDT makes no difference in case of thin shells <10 nm, leading to the conclusion that the vesicle thickness due to the correspondingly very thin vesicular shells (with respect to the length scale of the light scattering experiment) usually does not contribute to the *q* dependence of the particle form factor in the *q* range covered by light scattering.

Next, Pencer and Hallett considered the effect of vesicle polydispersity on measured particle form factors, assuming a Gamma function for the size distribution $G(R)$ in their simulations of the average particle form factor:

$$G(R) = \left(\frac{m+1}{R_0}\right)^{m+1} \frac{R^m}{m!} \exp\left(-\frac{R\cdot(m+1)}{R_0}\right) \tag{5.71}$$

with R_0 the average vesicle size, $\sigma^2 = R_0^{\,2}/(m+1)$ the variance of the size distribution, $\Delta^2 = 1/(m+1)$ the size polydispersity, and *m* an adjustable parameter.

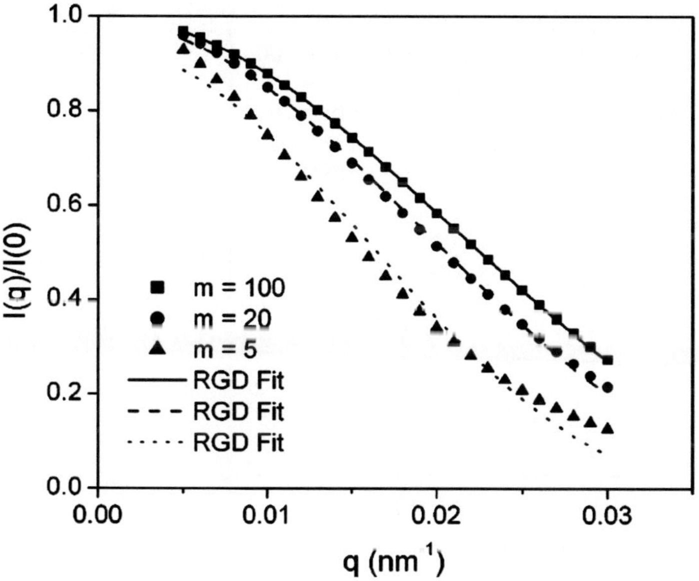

Fig. 5.37. Simulated particle form factors for polydisperse vesicles of average size 60 nm. Reprinted with permission from J. Pencer and F.R. Hallett, Langmuir 19, 7488–7497, 2003, Copyright 2003 American Chemical Society (ref. [5.19]).

The corresponding average particle form factor measured by static light scattering is given as a z-average according to:

$$\langle P(q)\rangle = \frac{\int_0^\infty M_R^2 P(q,R)G(R)dR}{\int_0^\infty M_R^2 G(R)dR} \tag{5.72}$$

with M_R the mass of vesicles of radius R. Simulated average form factors using this approach to describe polydisperse spherical vesicular systems are shown in Fig. 5.37.

Guinier fits to the data shown in Fig. 5.37 should yield the z-average radius of gyration $\langle R_g \rangle_z$. If such fits are attempted outside the acceptable range $qR<1$, the Guinier approach tends to overestimate the particle size for small size polydispersities and strongly underestimate it for larger size polydispersities. For the sample with the highest polydispersity ($m=5$), fits using the RGDT function derived for monodisperse vesicles show strong deviations at larger q values, that is, close to the minima of the particle form factor (see dotted line in Fig. 5.37). In conclusion, an accurate determination of $\langle R_g \rangle_z$ affords light scattering measurements at small scattering angles (small q), whereas effects of sample polydispersity are best detected around the vicinity of the minimum in $P(q)$, that is at very large q in case of the example regarded here.

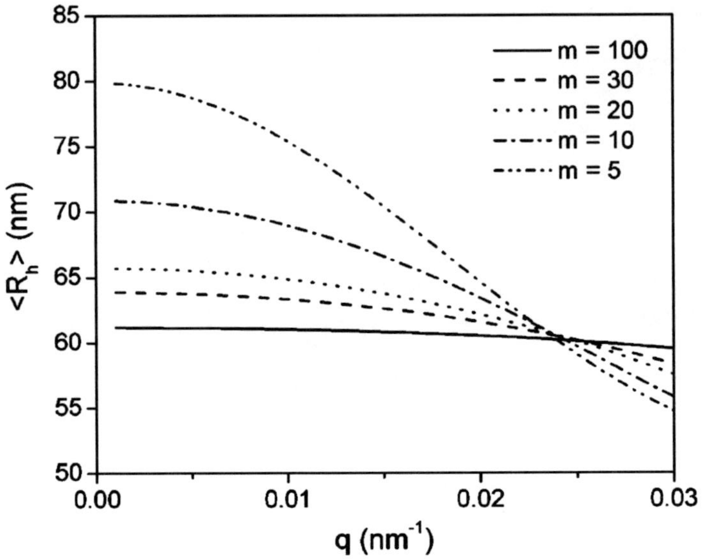

Fig. 5.38. Simulated apparent hydrodynamic radii for polydisperse vesicles of average size 60 nm. Reprinted with permission from J. Pencer and F.R. Hallett, Langmuir 19, 7488–7497, 2003, Copyright 2003 American Chemical Society (ref. [5.19]).

Polydispersity also has a major influence on the angular dependence of the apparent particle hydrodynamic radius detected in a dynamic light scattering experiment, as discussed in detail before. Simulated results obtained for the same polydisperse systems presented in Fig. 5.37 are shown in Fig. 5.38.

Finally, the topological shape of the vesicles also influences the form factor $P(q)$. So far, only spherical particles have been considered. In case of non-spherical ellipsoidal vesicles, the particle form factor in the limiting case of thin shells depends on the two symmetry axes *a,b* according to:

$$P(q,a,b)=\int_0^1\left[\frac{\sin(qu)}{qu}\right]^2 dx \text{ , and } u=\sqrt{a^2x^2+b^2\left(1-x^2\right)} \tag{5.73}$$

Here, $x=\cos\alpha$, with α being the angle between vesicle symmetry axis and the line which bisects the angle made between incident and scattered laser beam, also called bisectrix. Anisotropic symmetrical vesicles considered here have two different morphologies: prolate, where *a* is the long axis, or oblate, in which case *a* is the short axis of the ellipsoid. In Fig. 5.39, particle form factors simulated for vesicles with given surface area, and therefore identical particle mass corresponding to that of spherical vesicles of radius 60 nm, but with varying eccentricity (= ratio a/b), are shown for prolate and oblate morphologies.

As can be seen in Fig. 5.39, there are systematic deviations between the spherical vesicle form factor (solid line) and the data simulated for ellipsoidal particles. For oblate vesicles, these deviations are less obvious since they lie outside the measurable q range.

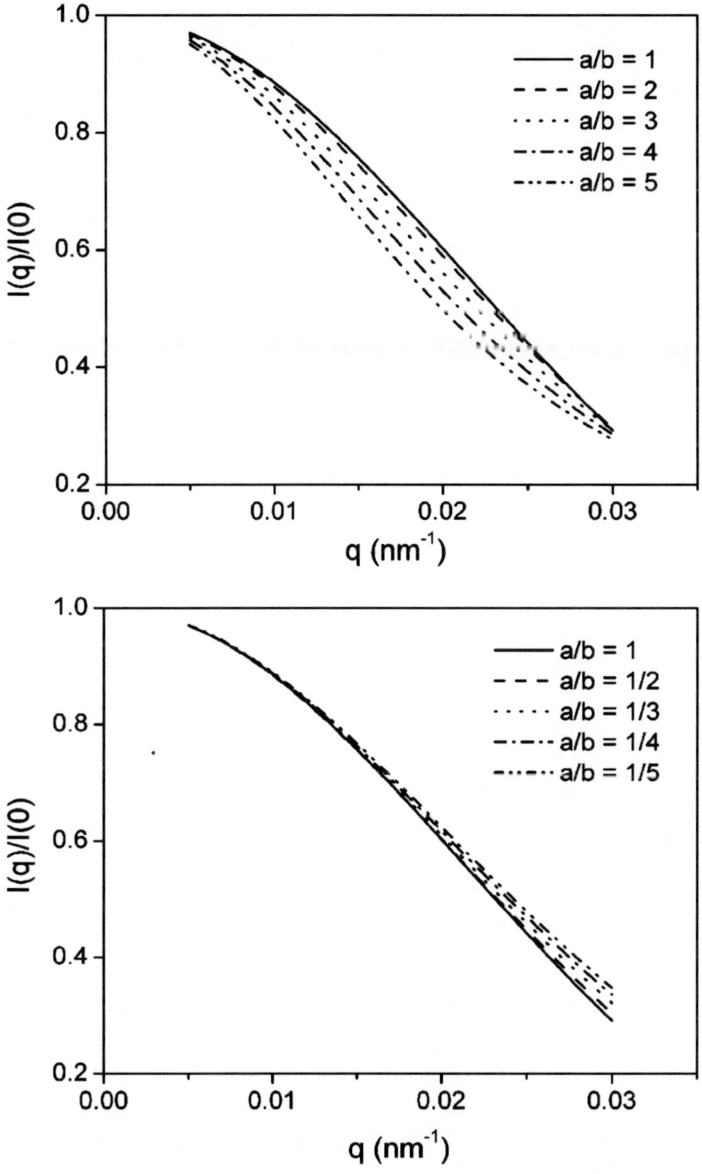

Fig. 5.39. Simulated particle form factors for ellipsoidal vesicles of identical surface area with prolate (*top*) and oblate (*bottom*) morphology. Reprinted with permission from J. Pencer and F.R. Hallett, Langmuir 19, 7488–7497, 2003, Copyright 2003 American Chemical Society (ref. [5.19]).

The above form factor of ellipsoidal particles can be rewritten as:

$$P(q,a,b) = \int_a^b G(R)\left[\frac{\sin(qR)}{qR}\right]^2 dR \text{ with } G(R) = \frac{1}{\sqrt{a^2-b^2}}\frac{R}{\sqrt{R^2-b^2}} \quad (5.74)$$

Interestingly, the expression given in Eq. 5.74 mathematically corresponds to that used for apparent particle form factors obtained from polydisperse spherical vesicles (Eq. 5.72). As a consequence, static light scattering from monodisperse ellipsoidal vesicles can formally be expressed in terms of scattering from polydisperse spherical vesicles. In experimental practice, it might be difficult or even impossible to deconvolute contributions from vesicle shape and size polydispersity using SLS data alone. As a consequence, it also becomes impossible to achieve a reliable characterization of the sample by static light scattering only. It should be noted that an ellipsoidal vesicle topology also has an effect on the hydrodynamic radius of the vesicles measured in a DLS experiment: whereas, compared to a spherical particle identical in molar mass, the hydrodynamic radius increases with increasing eccentricity for prolate particles, it decreases for oblate vesicles.

In conclusion, two different samples of polydisperse ellipsoidal vesicles exhibiting nearly identical particle form factors in a static light scattering measurement may yield different particle size distributions in a dynamic light scattering measurement. This opens up a strategy to characterize such a vesicular system, using a combination of SLS and DLS, an approach which has been verified by Pencer and Hallett in experimental practice using samples and setups described above. The procedure is as following: first, particle form factor and intensity-weighted size distribution are measured by static and dynamic light scattering for a given sample. Then, a particle shape (prolate, oblate, spherical) and plausible value for the axial ratio are assumed as a starting point for the data fitting. Next, these input parameters are used to convert the intensity-weighted size distribution measured by DLS to a number-weighted size distribution. Using this number-weighted size distribution as input along with the assumed particle shape and axial ratio, it is now possible to simulate the experimental particle form factor measured by static light scattering. This procedure is repeated until the best match between simulated and measured $P(q)$ is found.

(iv) Experimental results:
Figure 5.40 shows the intensity-weighted size distributions determined from the DLS data of both DOPC and SOPC vesicles.

Here, the reader should note that the size distributions shown in Fig. 5.40 had been calculated from DLS-data measured at finite scattering angle 90°. As discussed above in Chap. 1.3, this is not the "true" particle size distribution but only an apparent distribution.

Using nevertheless the size distribution shown in Fig. 5.40 as input, the authors simulated $P(q)$ for various particle morphologies, as shown in Fig. 5.41 for the example of DOPC vesicles.

In conclusion, the authors obtained an oblate vesicle shape with axial ratio smaller than $1/2.5$, which is a fairly large deviation from the originally expected spherical shape of extruded vesicles.

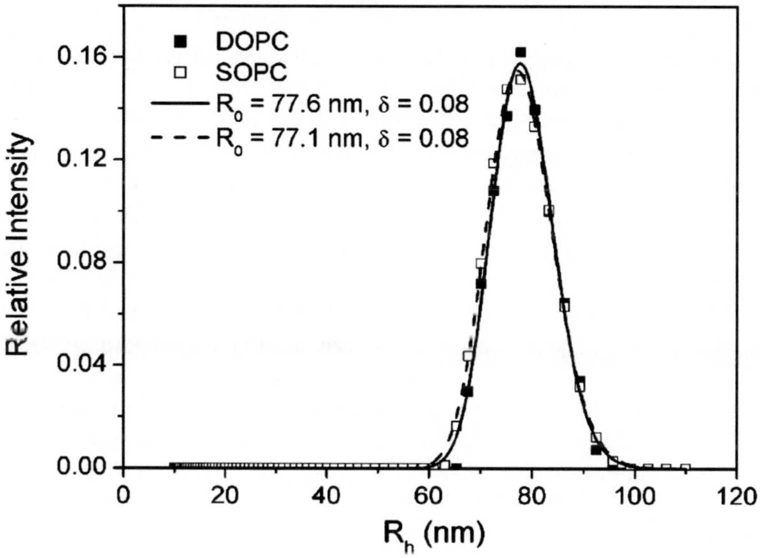

Fig. 5.40. Size distributions for unilamellar vesicles obtained from DLS measurements. Reprinted with permission from J. Pencer and F.R. Hallett, Langmuir 19, 7488–7497, 2003, Copyright 2003 American Chemical Society (ref. [5.19]).

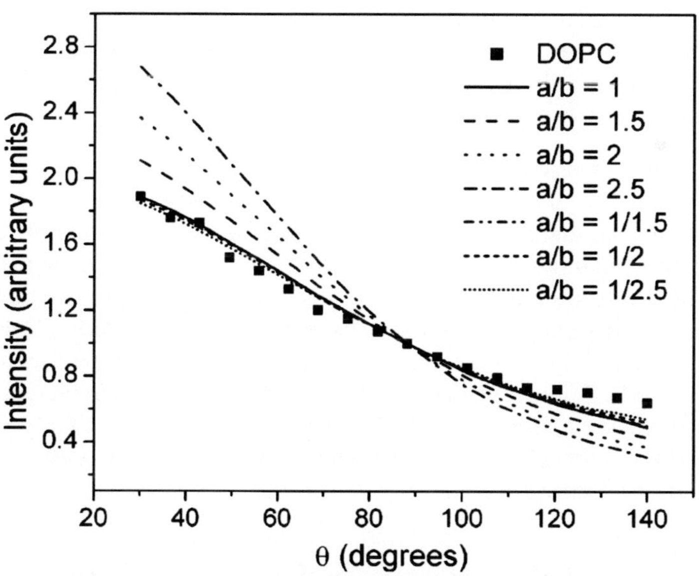

Fig. 5.41. Experimental scattered intensity obtained from static light scattering of extruded DOPC vesicles, and simulations using various particle morphologies. Reprinted with permission from J. Pencer and F.R. Hallett, Langmuir 19, 7488–7497, 2003, Copyright 2003 American Chemical Society (ref. [5.19]).

(v) Conclusions:
This example demonstrates how, in case both dynamic light scattering and static light scattering as a stand-alone technique would be insufficient to characterize the scattering particles due to the impossibility of separating effects of sample topology and sample polydispersity on the form factor $P(q)$, a clever combination of both methods prevails. However, a major criticism remains: first, the authors used an apparent size distribution determined by DLS-measurements at one scattering angle only as input for their procedure. The accurate method would have been to determine apparent size distributions from DLS-data measured at various scattering angles and calculate the "true" size distribution by extrapolation to zero scattering angle ($q=0$). In addition, one has to consider that particle size distributions obtained by inverse Laplace transformation of DLS data, due to some ambiguity of the mathematical procedure, do not necessarily represent the "true" particle size distribution. Finally, the authors in case of q-dependent DLS measurements could have cross-checked their procedure by simulating the q-dependence of the apparent diffusion coefficient, using the respective particle form factors and the size polydispersity as input parameters.

Example F (Ref. [5.20]):
Our second example for light scattering from vesicular systems deals with a different approach to separate effects of polydispersity from effects of the shape of the scattering particle on $P(q)$. Since the vesicles studied here were very large, the authors have been able to develop a light scattering apparatus suitable to study the particle form factor obtained by light scattered from one single vesicle, in which case size polydispersity naturally plays no role.

(i) Samples and sample treatment:
Poly (L-lysine-alt-terephtalic acid) microcapsules filled with water were synthesized by interfacial polycondensation and subsequently dispersed in distilled water. Buffer was added to obtain a constant ionic strength of 0.1 at the given pH of the solution, which was systematically varied since the microscopic vesicles studied here are a promising candidate for pH-sensitive release systems. The microcapsules were sucked up into a capillary cell which then was sealed with epoxy resin at both ends. Microcapsules prepared in this way had an average size of more than 20 μm and a membrane thickness in the order of 1 μm. For the scattering experiment, microcapsules located as close as possible to the center of the capillary [see Sect. (ii)] were selected.

(ii) Light scattering setup:
The unusual setup developed by Dobashi et al. for their small-angle static light scattering experiment is shown in Fig. 5.42.

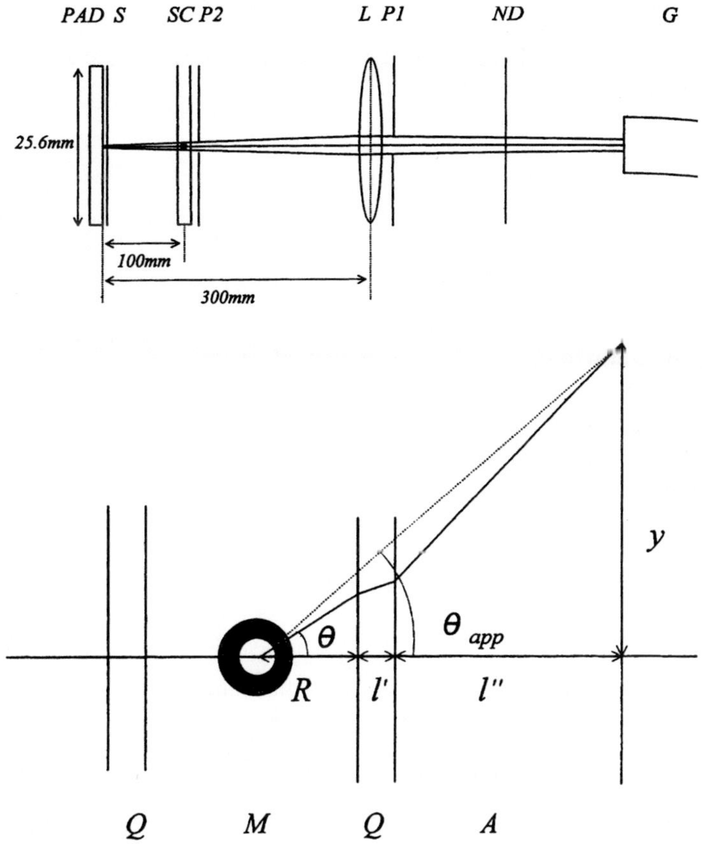

Fig. 5.42. Sketch of single particle small-angle static light scattering setup: PAD, photodiode array detector; S = Slit; SC = sample cell; P1, L; P2 = lens pinhole optics; G = HeNe laser (*top*); and relation between apparent and true scattering angle: *Q* = sample cell walls; M = microcapsule; A = air (*bottom*). Reprinted with permission from T. Dobashi, T. Narita, J. Masuda, K. Makino, T. Mogi, H. Ohshima, M. Takenaka and B. Chu, Langmuir 14, 745–749, 1998, Copyright 1998 American Chemical Society (ref. [5.20]).

Also shown is the relation between true and apparent scattering angle, since diffraction of scattered light especially at the silica-air-interface cannot be neglected here. Horizontally polarized light from the light source, a HeNe laser (5 mW, 632.8 nm), was transmitted through an optical pinhole setup and the sample cell onto a photodiode array detector (PAD) with 1,024 channels, having a pixel size of 25 μm × 1 mm for each channel (Hamamatsu S3904-1024Q). The smallest accessible q value of this setup was 0.5 μm^{-1}. By shifting the PAD in the vertical direction using a translational stage, the maximum q range could be extended up to a value of $q = 6\ \mu m^{-1}$. The sample cell was prepared from an inactivated fused silica capillary tubing with an inner diameter of 100 μm, thickness 138 μm, and length 5 cm.

(iii) Methods used for data analysis:
The form factor $P(q)$ of a spherical thin shell has been given in the last example. To adopt the terminology of ref. [5.20], we will repeat here the corresponding equations:

$$P(q)= \\ =(9\pi/2)\left[J_{3/2}(r_0 q)/(r_0 q)^{3/2}-(1-\delta/r_0)^3 J_{3/2}\left[(r_0-\delta)q\right]/\left[(r_0-\delta)q\right]^{3/2}\right]^2 \quad (5.75)$$

and $$J_{3/2}(rq)=\left[2/(\pi r^3 q^3)\right]^{1/2}\left[\sin(rq)-rq\cos(rq)\right] \quad (5.76)$$

r_0,δ are the outer radius and thickness of the spherical shell, and $J_{3/2}(rq)$ is the analytical function defined in Eq. 5.76. The major difference compared to the previous example E, where also light scattering characterization of vesicular particles is discussed, is that in case of example F the experimental qR range is sufficiently large to detect a couple of form factor oscillations, as will be shown further below. Therefore, exact fitting of the experimental data to the particle form factor (Eq. 5.75) using a least-squares-method yields the two parameters r_0,δ which characterize the spherical microcapsule. For infinitely thin shell thickness ($\delta\to 0$), the particle form factor of a hollow sphere is obtained:

$$P(q)=\left(\sin(r_0 q)/r_0 q\right)^2 \quad (5.77)$$

On the other hand, if the shell thickness reaches the particle size ($(r_0-\delta)\to 0$), one obtains the form factor of a compact sphere:

$$P(q)=\left(9\pi/(2r_0^3 q^3)\right)J_{3/2}^2(r_0 q) \quad (5.78)$$

In a double logarithmic plot, $\log P(q)$ vs. $\log q$, the asymptotic slope of the upper envelope at high q depends on the ratio of shell thickness δ to outer diameter $2r_0$. According to the corresponding equations, we expect a slope of –2 for hollow spheres (Eq. 5.77) and a slope of –4 for solid spheres (Eq. 5.78).

(iv) Experimental results:
Figure 5.43 shows both the measured particle form factor and a least-squares fit to the experimental data.

The inset in the figure indicates the slope of –2 expected for hollow spheres. The asymptotic slope of the experimental data corresponds well to the predictions for a hollow sphere, which is in agreement with the fact that the shell of the microcapsules is fairly thin compared to their size.

The parameters for the thin shell form factor fits to the SLS data, obtained from a single vesicle at different pH-values of the aqueous sample solution, as well as the particle size determined by optical microscopy are summarized in Table 5.10.

It should be noted that optical microscopy, although sufficient to determine the size of the capsules, is not capable of measuring the membrane thickness, underlining the necessity of the light scattering experiment in this case. Also, the experimental error of the vesicle size determined by microscopy is in the order of

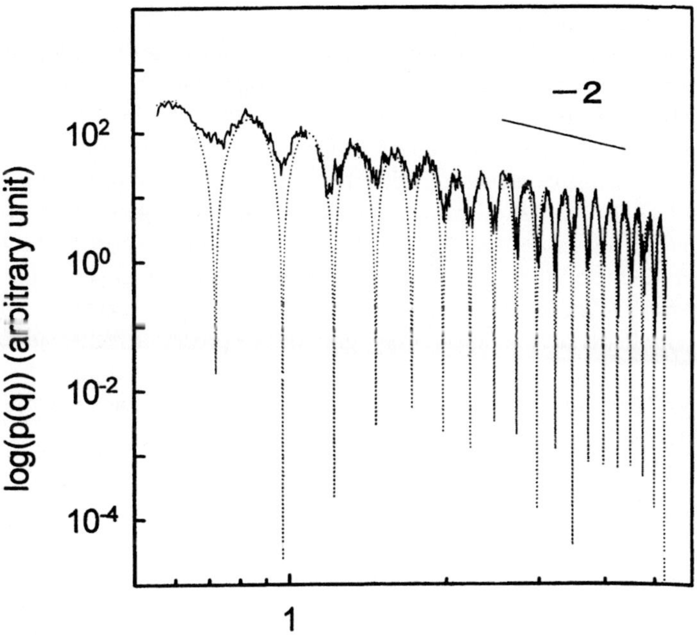

Fig. 5.43. Experimental and fitted (*dotted line*) particle form factor for a single microscopic vesicle. Reprinted with permission from T. Dobashi, T. Narita, J. Masuda, K. Makino, T. Mogi, H. Ohshima, M. Takenaka and B. Chu, Langmuir 14, 745–749, 1998, Copyright 1998 American Chemical Society (ref. [5.20]).

the resolution of an optical microscope (about 1 μm) and therefore is much larger than the error of the values determined by SLS. Finally, the pH of the sample solution seems to have a strong effect on capsule size and capsule wall thickness, both following the same trend: a decrease in overall capsule size corresponds to a decrease in membrane thickness and vice versa, indicating an isotropic volumetric phase transition of the hydrogel membrane of the studied microcapsules at pH 6.

Table 5.10. Sample characteristics for vesicles at different pH as determined from form factor fits, and average vesicle size measured by optical microscopy (OM). Reprinted with permission from T. Dobashi, T. Narita, J. Masuda, K. Makino, T. Mogi, H. Ohshima, M. Takenaka and B. Chu, Langmuir 14, 745–749, 1998, Copyright 1998 American Chemical Society (ref. [5.20]).

pH of medium	Membrane thickness δ/μm	Vesicle diameter d/μm by SLS	Vesicle diameter d/μm by OM
2	0.6 ± 0.1	29.4 ± 0.1	28 ± 1
4	0.2 ± 0.1	25.4 ± 0.1	24 ± 1
6	0.6 ± 0.3	56.8 ± 0.3	54 ± 3
8	1.2 ± 0.1	71.5 ± 0.2	71 ± 4
10	1.0 ± 0.1	68.9 ± 0.2	66 ± 3

(v) Conclusions:
We have seen how polydispersity effects can be avoided in static light scattering experiments if light scattered from one single particle is detected. This is only an option for large particles several micrometers in size, as studied here. Such objects in principle could also be investigated by ordinary optical microscopy, but microscopy sometimes cannot reveal certain structural details of the sample structure, as in case of the thin shell of the microcapsules investigated here. It has been shown by this example that light scattering can be a very important tool even for the detailed characterization of larger particles. Finally, it should be noted that single particle characterization, compared to the characterization of an ensemble of polydisperse particles, may give nonrepresentative results. Therefore, the single particle experiments described here should be repeated at least several hundred times for different individual particles to achieve a reasonable sample characterization.

Example G (Ref. [5.21]):
This example represents the experimental characterization of worm-like micelles by static and dynamic light scattering.

(i) Samples and sample treatment:
The worm-like micelles were formed in aqueous solution by association of the amphiphilic diblock copolymer poly-butadiene-b-poly(ethylenoxide) with block lengths 125 and 155, respectively. The polydispersity of the single copolymer chain was determined by gel permeation chromatography as $M_w/M_n = 1.02$. A stock solution of this copolymer of 0.2 wt% was prepared in high purity water from a Milli-*Q* purification system (Millipore/Waters), and diluted to five lower concentrations between 0.02 wt% and 0.12 wt%. All samples were directly filtered through cartridge filters with a membrane pore size of 0.2 μm (Schleicher & Schuell, Spartan 30/0.2 RRC, or Whatman anatop 10) into dust-free cylindrical quartz cells (Hellma, Germany) with an inner diameter of 0.8 cm.

(ii) Light scattering setup:
The LS setup used by Fuetterer et al. was a commercial machine for simultaneous static and dynamic light scattering from ALV. The light source was a Coherent 315M-150 frequency doubled diode pumped solid state laser, operating at 532 nm and 150 mW. Samples were immersed in a toluene bath for temperature control and suppression of unwanted refraction. The scattered light was detected with a photomultiplier tube (Thorn EMI) mounted on a goniometer arm. For DLS, time intensity correlation functions were calculated using an ALV-5000 multiple hardware correlator with 256 channels and first lag time 200 ns. Scattering angles during the DLS experiment were varied between 30° and 150°, corresponding to scattering vectors between 0.008 nm^{-1} and 0.030 nm^{-1}. For static light scattering, the scattered intensity was normalized to the primary beam intensity and brought to absolute scale using the solvent toluene as a scattering standard. Here, the primary beam was monitored employing a beam splitter and a four-segment photo-

diode. In addition, the refractive index increment of the sample, needed for SLS data evaluation, was measured with a commercial refractometer DR1/b from Baur Electronics. Like the DLS experiments, the SLS measurements were conducted at scattering angles between 30° and 150°.

(iii) Methods used for data analysis:
For elongated structures, several model form factors can be used to fit the dependence of measured scattered intensity on the scattering vector q. The simplest approach is that of monodisperse stiff thin cylinders with length L, where the form factor $P(q)$ is given as:

$$P(q)=\frac{2}{qL}\int_0^{qL}\frac{\sin(ql)}{ql}dql-4\left(\frac{\sin(qL/2)}{qL}\right)^2 \tag{5.79}$$

Self-assembled worm-like aggregates as considered here are usually polydisperse. For several comparable systems, the length distribution could be described with the asymmetric Schulz-Zimm distribution according to:

$$p(L)=\frac{\left[(k+1)/L_w\right]^{k+1}L^k\exp\left[-(k+1)L/L_w\right]}{\Gamma(k+1)} \tag{5.80}$$

with $k=1/(M_w/M_n-1)$ a polydispersity parameter, M_w/M_n the ratio of weight average molar mass to number average molar mass, $\Gamma(k+1)$ the Gamma function, and L_w the weight average contour length. This leads to the form factor of polydisperse stiff rods given as:

$$\langle P(q)\rangle=\frac{\int_0^\infty L\cdot p(L)P(q)dL}{\int_0^\infty L\cdot p(L)dL} \tag{5.81}$$

This expression (Eq. 5.81) corresponds to a mass-average particle form factor, as can be seen from the factor L (= particle contour length) corresponding to the particle mass.

Usually, worm-like micelles are not properly described as stiff cylinders since they retain a finite flexibility. This flexibility can be characterized by the ratio of the Kuhn-length L_K (which is the length of a stiff segment of a flexible aggregate chain) to the contour length. In this respect, $L/L_K=1$ corresponds to a stiff cylinder. For the q range covered in standard light scattering experiments, Koyama has derived a form factor for flexible worm-like chains [5.22] given as:

$$P(q)=\frac{1}{l_K^2}\int_0^{l_K}(l_K-x)\exp\left(-\frac{1}{3}q'^2\cdot xf(x)\right)\frac{\sin\left[q'\cdot xg(x)\right]}{q'\cdot xg(x)}dx \tag{5.82}$$

with the dimensionless parameter l_K related to the contour length L, $xf(x), xg(x)$ analytical functions depending on the Kuhn length L_K, and q' an analytical function depending on the scattering vector q (see ref. [5.22] for details). It should be noted that the approach by Koyama is only valid for cylinders of comparatively low flexibility.

The dynamic light scattering results were analyzed following the approach by Pecora to describe the amplitude correlation function for rod-like particles, as shown in a recent book on dynamic light scattering from rigid rods by Paul Russo [5.23]:

$$g_1(q,\tau) = \sum_{n=0}^{3} S_n(2n, qL) \cdot \exp\left(-\left\{D_T q^2 + \left[2n(2n+1)\right] D_R\right\}\tau\right) \tag{5.83}$$

D_T, D_R are the translational and rotational diffusion coefficients (see also Chap. 5.1). The amplitudes S_n depend on the length scale of the DLS experiment: for comparatively long diffusion distances corresponding to $qL \leq 4$, only pure translational diffusion (S_0) contributes to the correlation function, whereas for intermediate length scales $4 < qL < 15$ the first three relaxation modes ($n = 0, 1, 2$) have to be considered. The diffusion coefficients of rods of length L and diameter d can be described according to Kirkwood and Riseman [5.23] as:

$$D_T = \frac{kT}{3\pi\eta L} \ln\left(\frac{L}{d}\right) \quad , \quad D_R = \frac{9D_T}{L^2} \tag{5.84}$$

Fuetterer et al. claimed that the model of Broersma (see Chap. 5.1.) does not provide a better representation. Finally, the polydispersity of the cylindrical structures has also to be considered for interpretation of the DLS measurements, leading to a weight average amplitude correlation function:

$$\left\langle g_1(q,\tau)\right\rangle = \frac{\int_0^\infty L \cdot p(L) g_1(q,\tau) dL}{\int_0^\infty L \cdot p(L) dL} \tag{5.85}$$

(iv) Experimental results:
To analyze the particle form factor of their worm-like micelles measured by SLS, the authors have plotted their SLS data in the so-called Holtzer representation ($q \cdot (Kc/R)^{-1}$ vs. q) as shown in Fig. 5.44.

Here, the reader should note that the plateau value, in case of a Holtzer representation of the static light scattering data, directly corresponds to the mass per length of a rod-like scattering particle. Also shown in Fig. 5.44 are the simulated particle form factors for polydisperse stiff rods ($L_w = 389 nm, M_w/M_n = 1.2$) and polydisperse semiflexible, worm-like chains with large Kuhn length ($L_w = 380 nm, M_w/M_n = 2.0, L_K = 410 nm$). Both model fits (straight lines in Fig. 5.44) are nearly identical. In contrast, form factors calculated for polydisperse semiflexible,

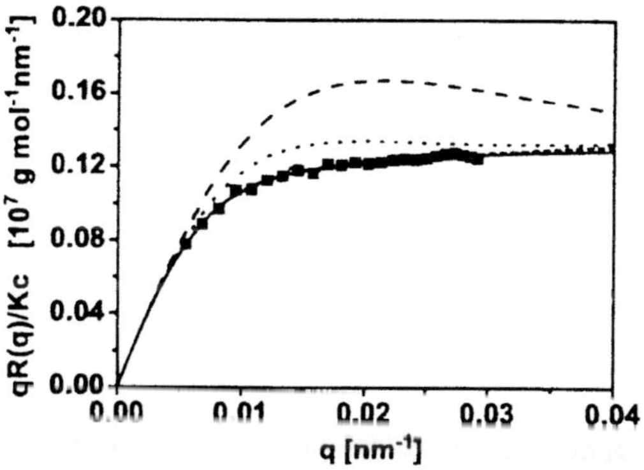

Fig. 5.44. Holtzer plot of SLS data obtained from wormlike copolymer micelles (c = 0.02 wt%). Reprinted with permission from T. Fuetterer, A. Nordskog, T. Hellweg, G.H. Findenegg, S. Foerster and C.D. Dewhurst, Physical Review E 70, 1–11, 2004, Copyright (2004) by the American Physical Society (ref. [5.21]). http://scitation.aip.org/getabs/servlet/GetabsServlet?prog=normal&id=PLEEE80000700000404140800000 1&idtype=cvips&gifs=yes. See text concerning the fitting functions also displayed in the figure.

worm-like chains assuming a higher flexibility or smaller Kuhn length (dashed ($L_K = 200nm$) and dotted ($L_K = 100nm$) line) cannot describe the data. This shows the very high stiffness of the cylindrical micelles studied here.

Next, Fuetterer et al. checked the validity of their fitting results by comparing the experimental radius of gyration (obtained from the SLS data by extrapolation to the small-q-regime using the Zimm equation) with values calculated from the fit parameters for polydisperse worm-like chains with Kuhn length L_K and weight average contour length L_w, using the following expression:

$$\left\langle R_g^{\,2}\right\rangle_z = L_K \frac{k+2}{6y} - \frac{L_K^{\,2}}{4} + \frac{L_K^{\,3}}{4L_w} - \frac{L_K^{\,4}}{8k(k+1)}\left(y^2 - \frac{y^{k+2}}{y+2/L_K}\right) \tag{5.86}$$

with $k = 1/(M_w/M_n - 1)$ and $y = (k+1)/L_w$. Alternatively, the radius of gyration for polydisperse rigid rods can be calculated as:

$$\left\langle R_g^{\,2}\right\rangle_z = \frac{(k+3)(k+2)}{(k+1)^2}\frac{L_w^{\,2}}{12} \tag{5.87}$$

Both the worm-like chain model and the rigid rod model reproduce the experimental result of $\left\langle R_g^{\,2}\right\rangle_z^{0.5} = 125nm$, assuming a molar mass mW of 1.67×10^8 g/Mol for the cylindrical aggregates. The later result also agrees extremely well with the value determined by Zimm analysis of the data ($M_w = 1.62\times10^8$ g/Mol). Since the investigated solutions were highly diluted, the SLS results were independent of sample concentration. From the molar mass, the weight-average number of copolymer chains per cylindrical aggregate was calculated as 12,100, corresponding to about 33 copolymer chains per nm length of the cylindrical micelle as determined from the contour length L_w.

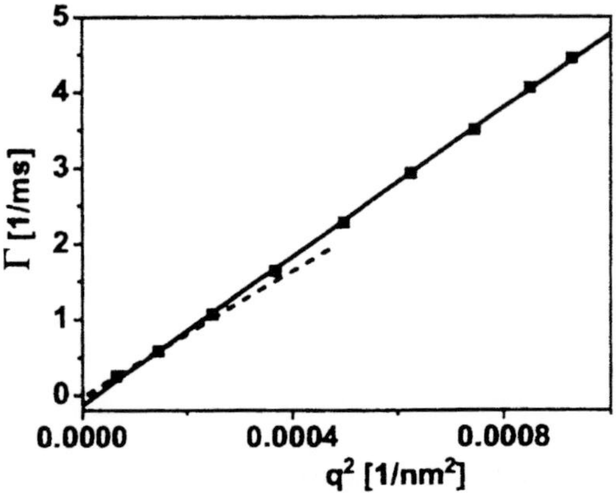

Fig. 5.45. Average decay rates obtained by CONTIN analysis of DLS results from cylindrical micelles (conc. 0.2 wt%), plotted versus q^2. Reprinted with permission from T. Fuetterer, A. Nordskog, T. Hellweg, G.H. Findenegg, S. Foerster and C.D. Dewhurst, Physical Review E 70, 1–11, 2004, Copyright (2004) by the American Physical Society (ref. [5.21]). http://scitation.aip.org/getabs/servlet/GetabsServlet?prog=normal&id =PLEEE8000070000004041408000001&idtype=cvips&gifs=yes

Dynamic light scattering data analyzed by the CONTIN method revealed only one significant relaxation mode. The dependence of the decay rate of this process on the scattering vector is shown in Fig. 5.45.

The dashed line shown in Fig. 5.45 corresponds to a linear fit to the decay rates at small q only, in which case the intercept corresponds to the expected value of 0. The linear fit over the whole q range, in contrast, leads to a significant deviation from the zero intercept, indicating the presence of additional relaxation processes or "higher modes," as expected for rod-like particles at higher q. The mean diffusion coefficient extracted from the low-q linear fit was $\langle D \rangle_z = 4.0 \pm 0.4 \cdot 10^{-2}\, nm^2 s^{-1}$, corresponding to a hydrodynamic particle radius $R_H = 60 \pm 6 nm$. From R_g and R_H, the authors calculated a ρ-ratio $R_g / R_H \approx 2$, a value to be expected for cylindrical scattering particles.

(v) Conclusions:

We have seen here an example for the characterization of polydisperse cylindrical objects in solution by light scattering. Importantly, the form factor fitting models have been checked for plausibility of the obtained results by comparison with theoretical predictions, using the experimental SLS results obtained by the Zimm approach in the low-q limit, and therefore independent of the fitting models, as input parameters. It should be noted that the Holtzer plot of the SLS data shown in Fig. 5.44 displays the plateau behavior expected for static light scattering from rod-like particles only over a comparatively narrow q-regime. In this case, the reliability of the resulting mass per length obtained from the plateau value of the Holtzer representation may be questionable. Additional sample characteristics like the diameter of the cylinders could be determined by neutron scattering, providing access to the particle form factor at a much higher

q-regime, corresponding to much smaller length scales than investigated by light scattering.

Example H (Ref. [5.24]):
The second example of light scattering from semiflexible rods shows a combination of fractionation (by GPC) and SLS to avoid polydispersity contributions to the light scattering signal, thereby allowing a less complicate and also more accurate data analysis compared to the previous example G. However, not all samples can be fractionated by GPC as described here: for self-associated cylindrical micelles in aqueous solution as treated in example G, one would expect that the structure would be seriously influenced if the sample would be squeezed through a GPC column at high pressure.

(i) Samples and sample treatment:
So-called polystyrene polymacromonomers were synthesized by radical polymerization of anionically prepared methacryloyl end-functionalized polystyrene macromonomers, leading to so-called bottle-brushes with the topology of semiflexible cylinders. Polymacromonomers with different side chain lengths or molar masses of the precursor macromonomers, in the range from 720 to 4,940 g/Mol with polydispersity $M_w/M_n < 1.05$ as characterized by MALDI-TOF mass spectrometry, were investigated.

(ii) Light scattering setup:
DLS measurements were performed with a standard ALV SP-86 goniometer, equipped either with an ALV3000 linear-τ hardware correlator and a Kr^+ ion laser (647.1 nm, 500 mW), or an ALV5000 multi-τ hardware correlator and a NdYAG laser (532.0 nm, 80 mW). Correlation functions were analyzed using the Cumulant method available within the commercially provided software package.

The SLS measurements were conducted in combination with GPC to obtain the characteristics of monodisperse sample fractions. The GPC setup was connected to an on-line Knauer combined viscometer/refractive index (RI) detector and an ALV1800 multiangle light scattering instrument (19 angles plus one monitor channel), equipped with an Ar^+ ion laser (514 nm, 300 mW) and a home-made cylindrical flow cell with 38 μL total volume. Typical runs used a 100 μL injection volume of the polymacromonomer toluene solution, with sample concentration 2–3 g/L. Absolute calibration of the SLS setup was achieved with toluene as a scattering standard in a cylindrical Suprasil glass cuvette, which then was replaced by the cylindrical flow cell also consisting of Suprasil glass, with 6-mm inner diameter, 20-mm outer diameter and 1-mm height. The scattering intensity of the pure flowing solvent toluene was recorded, and this value was later subtracted from the scattering intensity measured from the eluting polymacromonomer solution to obtain the excess scattering intensity. Typically, the scattering intensity was simultaneously detected at all 19 scattering angles over a period of 2 s at time intervals of 4 s during elution of the sample.

(iii) Methods used for data analysis:
SLS data obtained from the individual sample fractions were analyzed with the Zimm model, yielding the radius of gyration R_g and the molar mass M_w. To obtain more detailed information on the sample topology, the dependence of R_g on M_w was analyzed according to predictions by the worm-like chain model:

$$R_g^2 = \frac{LL_k}{6} - \frac{L_k^2}{4} + \frac{L_k^3}{4L} - \frac{L_k^4}{8L^2}\left(1 - \exp\left(-2 L/L_k\right)\right) \tag{143}$$

with L the contour length and L_k the Kuhn length (see also example G). To fit the experimental data to this equation, the contour length L has to be known precisely. Here, L was determined from the molar mass of the side chains and M_W of the polymer brush, assuming a monomer unit length of 0.25 nm, which corresponds to a vinyl main chain. Whereas this approach is certainly correct for linear polymer chains, it might lead to wrong results for polymer brush systems as treated in this example.

(iv) Experimental results:
As an example, Fig. 5.46 shows the reduced scattered intensity plotted versus q^2 for three sample fractions.

The data obtained for the high molar mass fractions (Fig. 5.46a) were measured with high accuracy despite the low sample concentration of $6.22 \cdot 10^{-3}$ g/L (as determined with the online RI detector of the GPC-MALS setup). For the small particle fraction, the data show significant scatter, but still allow a reasonable accurate determination of $\langle R_g \rangle_z$ and M_w. The curves shown in Fig. 5.46 correspond to the following sample characteristics:

(a) $R_g = 100.6 nm, M_w = 1.01 \cdot 10^7 \, g/mol, c = 6.22 \cdot 10^{-3} \, g/L$,

(b) $R_g = 47.7 nm, M_w = 3.77 \cdot 10^6 \, g/mol, c = 4.21 \cdot 10^{-2} \, g/L$, and

(c) $R_g = 14.3 nm, M_w = 7.61 \cdot 10^5 \, g/mol, c = 4.14 \cdot 10^{-2} \, g/L$.

Figure 5.47 shows the dependence of the radius of gyration on molar mass obtained by GPC-MALS from one exemplary polydisperse polymacromonomer bottle brush sample, including fits according to the worm-like chain model assuming two different values for the monomer chain length [l = 0.25 nm (straight line) and l = 0.22 nm (dotted line)].

Importantly, both fits shown in Fig. 5.47 agree very well with the experimental data, and it is impossible to determine the two fit parameters l and L_k in an unambiguous way from the SLS results.

To preserve good agreement between fitting function and experimental data, a decrease of the L_k values has to be compensated by an increase of the monomer length l and therefore the contour length L. This is to be expected, since at given molar mass the overall particle size, corresponding to the radius of gyration

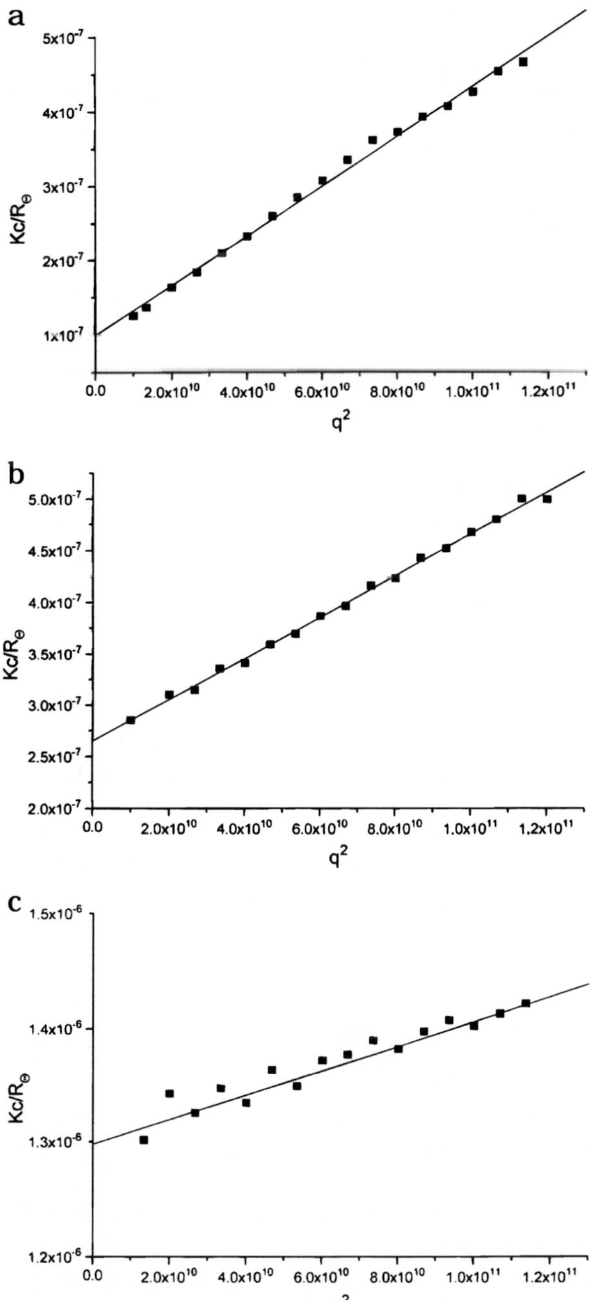

Fig. 5.46. Online measured reduced scattering intensity plotted versus q^2, for sample fractions obtained by GPC of a polydisperse polymacromonomer sample at different elution times. Reprinted with permission from M. Wintermantel, M. Gerle, K. Fischer, M. Schmidt, I. Wataoka, H. Urakawa, K. Kajiwara and Y. Tsukahara, Macromolecules 29, 978–983, 1996, Copyright 1996 American Chemical Society (ref. [5.24]).

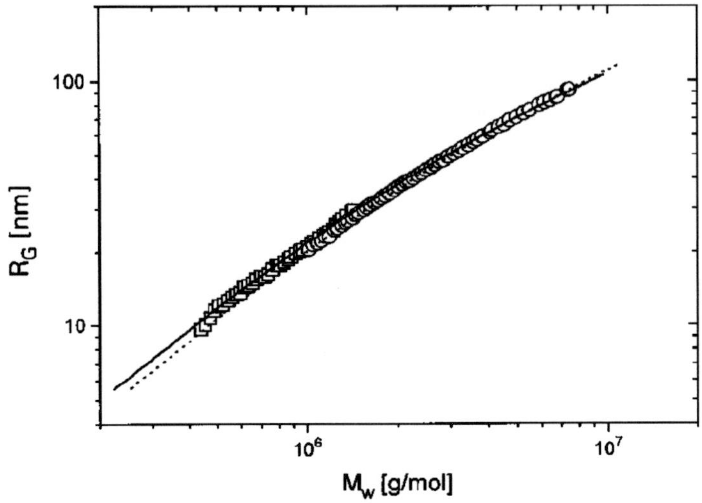

Fig. 5.47. Dependence of radius of gyration R_g on molar mass M_w for a polymacromonomer brush with side chain molar mass 2,900 g/mol including data fits using the worm-like chain model. Reprinted with permission from M. Wintermantel, M. Gerle, K. Fischer, M. Schmidt, I. Wataoka, H. Urakawa, K. Kajiwara and Y. Tsukahara, Macromolecules 29, 978–983, 1996, Copyright 1996 American Chemical Society (ref. [5.24]).

$R_g(M)$, has to remain unchanged. Restricting the discussion to the chain flexibility (defined by the L_k values) obtained for a fixed value $l = 0.25$ nm, Wintermantel et al. found an increase in chain stiffness with increasing length of the side chains of the bottlebrushes (from $L_k = 89nm$ for polymacromonomers with side chain molar mass 2,900 g/Mol, to $L_k = 207.6nm$ for side chains of 5,000 g/Mol). According to ref. [5.24], the systematic deviations between worm-like chain fits and experimental data in the small-M_w-regime, as visible in Fig. 5.47, increased with increasing side chain length. This was attributed by the authors to nonnegligible contributions of the cross-sectional diameter of the bottlebrushes to the radius of gyration if the brush thickness gets comparable to its contour length, an effect not considered in the worm-like chain model.

(v) Conclusions:

This example illustrates how fractionation can be used to provide an accurate analysis of the particle form factor in case of a comparatively complicated sample topology in combination with sample polydispersity. The merits of the GPC fractionation shown here are not only given by simply making the analysis of the detected light scattering signals easier due to elimination of polydispersity effects. Interestingly and more importantly, in case of very polydisperse samples the GPC-MALS coupling provides the experimental means of analyzing the particle morphology by the way of its characteristic dependence of particle size on molar mass over a wide continuous range of molar masses. It should be noted that, in analogy to example G, recently small angle neutron scattering (SANS) with its much smaller experimental length scale (compared to light scattering) has been used to explore the structure of polymacromonomer bottle brushes in more detail than possible from only light scattering characterization.

Example I (Ref. [5.25]):

Our next example deals with the thermo-responsive linear polymer poly-N-isopropylacrylamide (PNIPAM), which in aqueous solution shows a topological phase transition from an expanded polymer coil to a collapsed globular state at temperatures of about 34°C. For the in situ study of the structural change of single PNIPAM chains during this phase transition, light scattering is one of the best-suited experimental methods, as will be illustrated by example I.

(i) Samples and sample treatment:

A stock solution of self-made high molar mass PNIPAM chains in deuterated water with a concentration of 2.5e-5 g/mL was prepared and kept at room temperature for 1 week to assure complete dissolution. This solution was further diluted in the range 2.5e-5 to 5.1e-7 g/mL and filtered through a 0.5-μm Millipore Millex-LCR filter. Samples were studied by light scattering at different temperatures after they had been equilibrated at the respective sample temperatures for at least 2 h.

(ii) Light scattering setup:

The setup used for both static and dynamic light scattering had an unusually small scattering angle range of 6° to 20°, making it very suitable to study long coiled polymers in solution since it fulfils the condition $qR_g \ll 1$, which is necessary to apply Zimm analysis to static light scattering data. More details are given in ref. [5.26].

(iii) Methods used for data analysis:

Static light scattering results were analyzed in the standard way using the Zimm equation, an approach which was valid throughout the experimental q range of the experimental small angle scattering setup used here. To obtain the hydrodynamic particle radius from dynamic light scattering, the Cumulant method in combination with the Stokes–Einstein–equation was applied, whereas the particle size distribution was determined by Laplace inversion of the correlation functions using the CONTIN algorithm.

(iv) Experimental results:

Figure 5.48 shows the results from SLS (Kc/R) and DLS ($f(R_H)$) measurements obtained at two different sample temperatures.

Both the decreasing slope in the q dependence of the average scattered intensity with increasing temperature and the corresponding decrease in hydrodynamic radius prove the strong shrinkage of the scattering polymer chains with temperature increase in a very narrow temperature interval. The experimental light scattering results shown in Fig. 5.48 correspond to a decrease in radius of gyration from 130 nm to 20 nm, and in hydrodynamic radius from 100 nm to

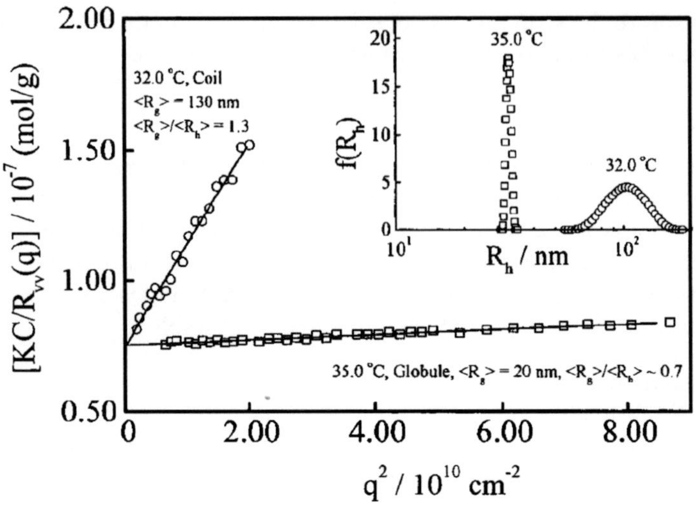

Fig. 5.48. Angular dependence of normalized inverse scattered intensity and hydrodynamic particle radius (*inset*) of PNIPAM chains in aqueous solution at two different sample temperatures. Reprinted with permission from X.H. Wang and C. Wu, Macromolecules 32, 4299–4301, 1999, Copyright 1999 American Chemical Society (ref. [5.25]).

29 nm. These values show that also the ratio $\rho = R_g / R_H$, and therefore not only the particle size but also the particle topology itself, is changing.

The intercept of the SLS data for $q \to 0$ is not depending on sample temperature, showing that the molar mass of the scattering particles remains unchanged, and that the change in particle size and shape with increasing temperature is a single-particle phenomenon. The second Virial coefficient of this system, as determined from a Zimm-analysis of the SLS results, is linearly decreasing with increasing temperature, reaching the value 0 (corresponding to the so-called Θ-state where the polymer coil resumes unperturbed dimensions) at the transition temperature T = 32°C. This shows a systematic decrease in solvent quality with increasing temperature, leading to the structural change from an expanded swollen polymer coil in a good solvent (low T) to a collapsed globular structure in a bad solvent (high T). This change in particle morphology is deduced from the values of $\rho = R_g / R_H$, which is 1.3 and 0.66 at T = 32°C and T = 35°C, respectively.

Figure 5.49 summarizes the experimental results in showing the radius of gyration and $\rho = R_g / R_H$ over the whole temperature range of the light scattering experiments.

Also shown in Fig. 5.49 is the effect of the solvent (by replacing D_2O with H_2O) (top), and the difference in the transition behavior upon heating or cooling of the sample (bottom). In all cases, the transition took place in a very narrow temperature regime of a few degrees close to the human body temperature, making PNIPAM a promising candidate for future controlled release architectures in drug delivery. In H_2O, the transition was slightly shifted to a lower

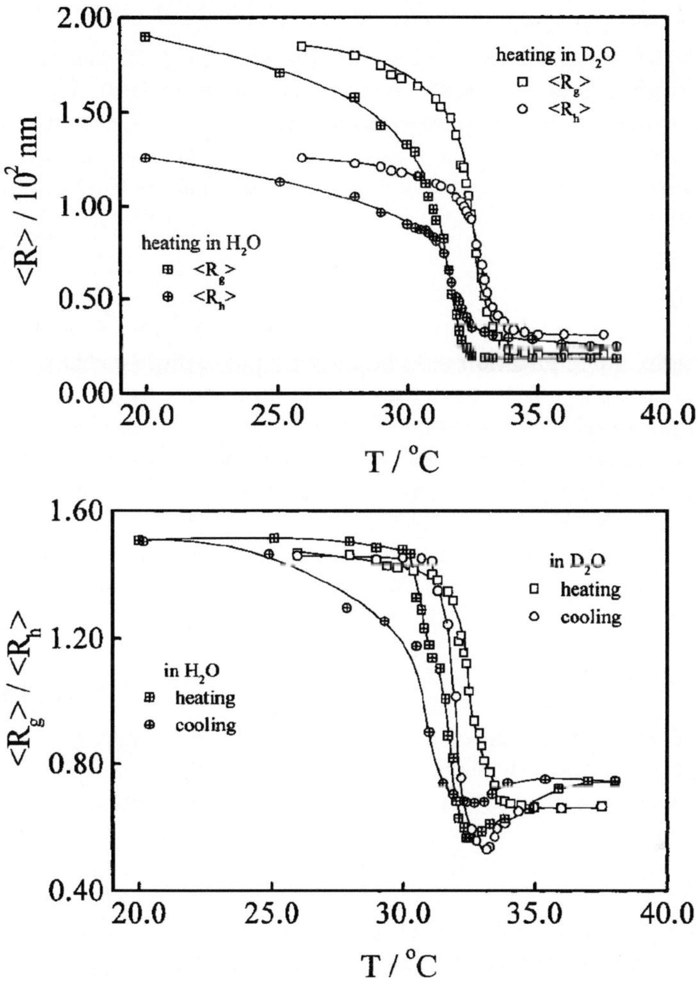

Fig. 5.49. Radius of gyration and hydrodynamic radius of PNIPAM for heating in different solvents H_2O and D_2O (*top*), and ρ-ratio for heating and cooling in H_2O and D_2O (*bottom*) vs. sample temperature. Reprinted with permission from X.H. Wang and C. Wu, Macromolecules 32, 4299–4301, 1999, Copyright 1999 American Chemical Society (ref. [5.25]).

temperature, indicating that D_2O is the better solvent for PNIPAM. Comparing the results obtained by heating or cooling, the transition temperature was nearly unchanged, whereas the ρ-ratio showed a pronounced minimum for the experimental curves obtained by cooling the sample. At this minimum, the ρ-ratio had an unusually small value of only 0.54, which according to Wang and Wu corresponds to a so-called molten globule state, where the surface of the spherical particle has a lower density than its center. This unusual (in comparison to a compact sphere structure with homogeneous density) topology lead to a decrease in R_g, while R_H remained unchanged, a scenario which provides a possible explanation for the unusually small value of $\rho = R_g / R_H = 0.54$.

(v) Conclusions:
We have reviewed here one more illustrative example how the combination of dynamic and static light scattering allows one to determine the morphology of scattering particles in solution in the nanoscopic size regime. In case a standard topology as coil or sphere is to be expected for the scattering particles, even the most simple methods used for analyzing light scattering data, such as Zimm model and Cumulant analysis, can be sufficient to characterize the system.

Example J (Ref. [5.27]):
This example for the experimental practice of static light scattering illustrates, like the preceding example, the application of light scattering to explore a change in particle structure by changing the solvent quality. This time, not a single-particle transition as with the coil-globule transition of PNIPAM chains (example I), but the aggregation of single spherical colloidal particles to larger fractal aggregates upon addition of salt to an aqueous solution is reviewed. Here, light scattering not only provides a characterization of the increase in average particle size, but also can be used to determine the size polydispersity of the growing aggregates and, more importantly, their fractal dimension, which provides insight in the aggregation mechanism itself.

(i) Samples and sample treatment:
The samples were self-made polystyrene latex particles prepared by the standard surfactant-free emulsion polymerization method using ammonium persulfate as radical initiator. After synthesis, the latex dispersion was dialyzed against Millipore water to remove residual monomer, ammonium persulfate, and any other ionic impurities. The particles were characterized by transmission electron microscopy, yielding an average particle diameter of 330 nm with a size polydispersity below 3%. It should be noted that, as a consequence of the initiator forming part of the latex particles, these colloidal spheres are highly charged in aqueous solution and thereby form a stable dispersion of single particles. If the Coulomb repulsion originating from these particle charges is screened by adding salt to the aqueous solution, the dispersion becomes unstable and the colloidal spheres form aggregates. To characterize the structure of these aggregates by light scattering, the required amount of polystyrene latex particles was added in the form of a salt-free aqueous dispersion to a solution of the salt KNO_3 of known concentration at room temperature, and this sample was fed into the light scattering cuvette immediately after mixing.

(ii) Light scattering setup:
Small-angle static light scattering measurements were performed using the commercial Malvern Mastersizer S equipped with a HeNe laser as light source. This multiangle setup (MALS) not only provides a simultaneous measure for the average scattered intensity at a range of scattering angles, but also yields the average particle size and size distribution of scattering particles within the sample determined from dynamic light scattering as a direct experimental output.

(iii) Methods used for data analysis:
By plotting the average scattered intensity versus the scattering vector, Burns et al. determined the fractal dimension d_f of the formed aggregates using the following relation (see Chap. 1.2):

$$\log I(q) = -d_f \log q \tag{5.89}$$

Note that this relation is only valid for the limited q range $1/R \ll q \ll 1/r_0$, with R the size of the aggregates, and r_0 that of the single spherical particles.

(iv) Experimental results:
Figure 5.50 shows typical experimental static light scattering data, from which the fractal dimension d_f of the colloidal aggregates was determined as indicated by the solid line.

For a given sample, scattered intensity and also fractal dimension were changing with time due to proceeding particle aggregation. To obtain the structure of the aggregates in "equilibrium," the value obtained for d_f in the plateau region was chosen, as shown in Fig. 5.51.

The experimental result that the fractal dimension of the clusters remained constant after a certain time, while the cluster size still was increasing as seen in Fig. 5.51, can be explained by a transition in the aggregation mechanism itself. Here, a more detailed discussion is beyond this book.

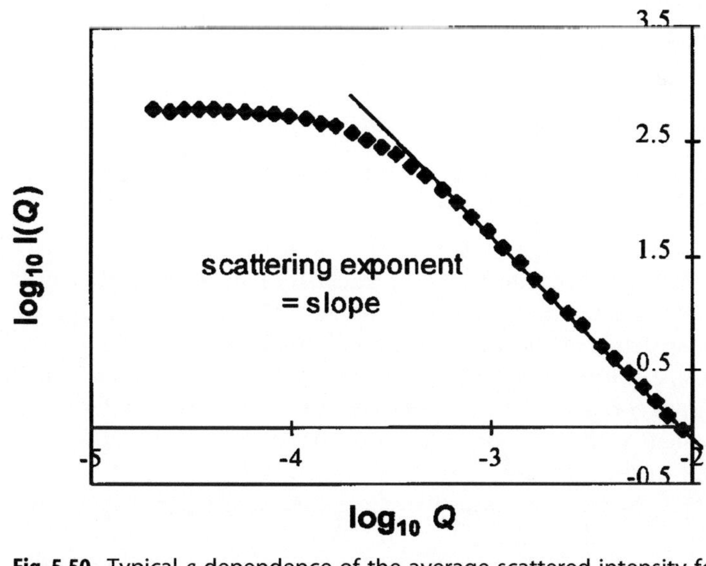

Fig. 5.50. Typical q-dependence of the average scattered intensity for aggregated colloidal particles (in double-logarithmic scale), from which the fractal dimension corresponding to the slope, as indicated, can be determined (scattering vector $Q = q$ measured in nm^{-1}). Reprinted with permission from J.L. Burns, Y.D. Yan, G.J. Jameson and S. Biggs, Langmuir 13, 6413–6420, 1997, Copyright 1997 American Chemical Society (ref. [5.27]).

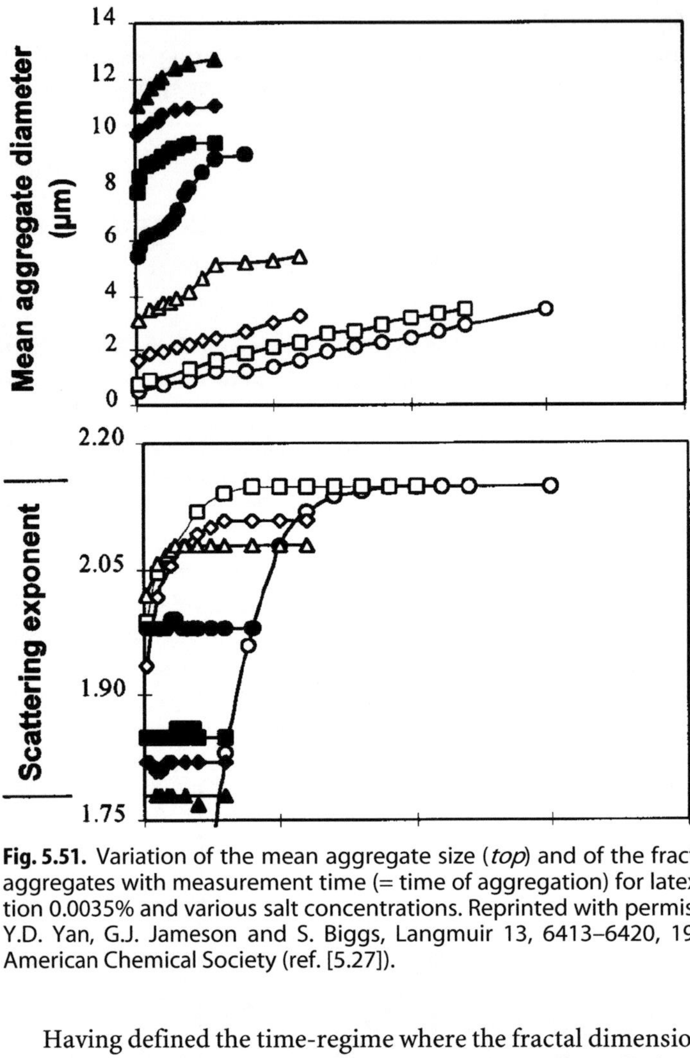

Fig. 5.51. Variation of the mean aggregate size (*top*) and of the fractal dimension of the aggregates with measurement time (= time of aggregation) for latex particle concentration 0.0035% and various salt concentrations. Reprinted with permission from J.L. Burns, Y.D. Yan, G.J. Jameson and S. Biggs, Langmuir 13, 6413–6420, 1997, Copyright 1997 American Chemical Society (ref. [5.27]).

Having defined the time-regime where the fractal dimension measured by SLS does not depend on measurement time and therefore all clusters, although continuously growing with increasing time, remain self similar in structure (= plateau regime), Burns et al. explored the effect of salt concentration on the fractal dimension of the growing aggregates. The values of d_f varied between 1.78 and 2.20, which is in agreement with literature values for diffusion-limited and reaction-limited cluster growth, respectively.

Here, just a brief comment on these two mechanisms of particle aggregation should be given: the fractal dimension d_f provides a measure for the compactness of a given 3D-structure: the lower the value of d_f compared to the limit for compact objects ($d_f = 3$), the more porous the structure. Clusters formed by diffusion limited aggregation are less compact since particles stick together at their random collision positions, whereas reaction limited aggregation allows for structural rearrangements after the particles have collided until neighboring particles are firmly stuck, leading to a cluster structure of higher compactness. As a consequence, more compact clusters were formed at lower salt concentration, where the latex dispersion was less unstable and consequently more collisions were needed for particle aggregation. On the other hand, at high salt concentration the Coulomb repulsion between the latex particles was perfectly screened, and any collision of freely diffusing latex particles and/or small particle clusters immediately resulted in further particle aggregation.

(v) Conclusions:
This example has shown us the use of small angle static light scattering to explore solutions of fractal aggregates, and also represents an experimental application for the concept of fractal dimensions introduced in Chap. 1.2 of this book.

Example K (Ref. [5.28]):
Our final example in this Chap. 5.2 about the investigation of particle topologies by light scattering describes a very important recent approach, which enables extraction of complicated density profiles of spherical scattering particles from SLS data. For this purpose, Schnablegger and Glatter have developed the so-called ORT program more than 10 years ago [5.29]. ORT is a computer-based program which numerically performs a Fourier transform of the measured average scattered intensity $I(q)$, and thereby transfers SLS curves into pair-distance distribution functions which can be converted into density profiles of the scattering particles, if a spherical particle topology is assumed. It should be stressed that the group of Glatter has a long tradition and high experience in numerically obtaining radial densities of particles from corresponding static scattering data. Their early calculations have been extended to very polydisperse and even interacting isotropic scattering particles.

(i) Samples and sample treatment:
Peytcheva et al. mixed 0.5 M aqueous $CaCl_2$ solution and 0.3 M aqueous Na_2HPO_4 solution in a so-called double-jet setup, adding the two solutions with a speed of 10 mL/h via thin capillaries first into a mixed jet and subsequently into a solution of the synthetic polypeptide poly(sodium)aspartate. In the mixed jet small calcium phosphate nuclei are immediately formed. If these nuclei are brought into contact with the polypeptide, stable aqueous dispersions of larger hydroxyapatite (HAP) colloidal nanoparticles are formed by Ostwald ripening. To study the effect of the polypeptide, its concentration was varied systematically. The dispersions were then filtered with Sartorius membrane filters of pore size 0.8 μm and introduced into the light scattering setup.

(ii) Light scattering setup:
The light scattering setup was a commercial apparatus (ALV) with ALV5000 multiple τ hardware correlator and a frequency-doubled Nd-YAG laser operating at 532 nm as the light source.

(iii) Methods used for data analysis:
Peytcheva et al. determined the average hydrodynamic radii of the growing nanoparticles by DLS, but no details of the data treatment have been given. Importantly, the internal structure or radial density profile of the colloidal HAP particles was determined from SLS data using the ORT program, which transforms the measured average scattered intensity $I(q)$ into pair-distance distribution functions (PDDF). Another program called DECON (for "deconvolution") was used to transform these PDDFs into the radial polarizability densities of the nanoparticles [5.29].

(iv) Experimental results:
Figure 5.52 shows the increase of the nanoparticles by Ostwald ripening for a solution containing 0.4 wt% polypeptide (M = 18,000 g/Mol).

The nuclei have a hydrodynamic radius of about 100 nm and, after an induction period of 10 min, grow to about 400 nm within 1 h. The results shown in Fig. 5.52 seem to be rather "noisy" especially at larger particle size, This is unexpected, since the accuracy of hydrodynamic radii determined by DLS usually is better than +/–10 nm in the size regime considered here, whereas the scattering of the data displayed in Fig. 5.52 reaches more than +/–50 nm in some cases. Since Peytcheva et al. did not provide any details how they determined their average hydrodynamic radii in ref. [5.28], for example, the q-dependence of the

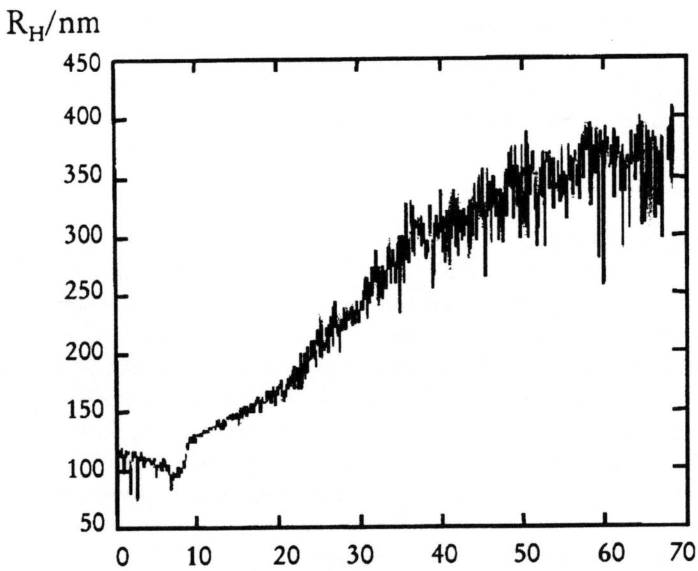

Fig. 5.52. Dynamics of structural growth of calcium phosphate colloids expressed by the average hydrodynamic radius measured by DLS. Reprinted with permission from A. Peytcheva, H. Cölfen, H. Schnablegger and M. Antonietti, Colloid Polym.Sci. 280, 218–227, 2002, Copyright 2002 Springer (ref. [5.28]).

apparent diffusion coefficients, etc., we refrain from speculating on the origin of this comparatively large DLS data noise.

The evolution of the average scattered intensity over 16 days for a sample containing 0.1 wt% polypeptide is shown in Fig. 5.53.

Although the average particle size as characterized by the hydrodynamic radius remained constant after a growth time of about 70 h (see Fig. 5.52), the particle morphology itself keeps changing for much longer times, as obvious from the change in the particle form factors displayed in Fig. 5.53. To investigate this effect in more detail, the authors have determined the radial density profiles of the particles as described in Sect. (iii). These results are shown in Fig. 5.54.

At the beginning, the particle density is lower in the center of the spherical nanoparticles than at their outer regions, corresponding to the morphology of a hollow sphere. With increasing time, the density in the particle center strongly increases, while the overall particle size remains nearly constant (about 500 nm). This rearrangement of the aggregate structure was attributed by Peytcheva et al. to a free rearrangement of crystal platelets from a kinetic structure called "hollow snowball" to a thermodynamically more stable structure with highest density at the particle center, therefore called "compact snowball". According to X-ray scattering and atomic force microscopy, the size of the crystal platelets was about 6 nm × 24 nm.

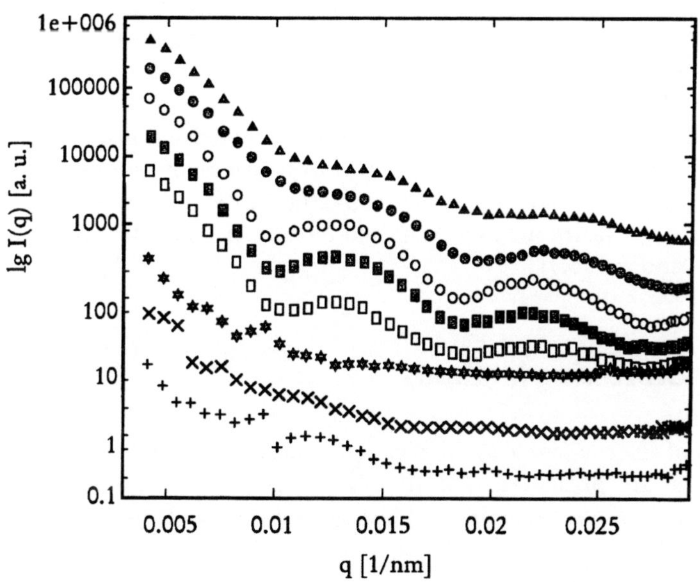

Fig. 5.53. Time-dependent static light scattering average intensities, describing the growth of calcium phosphate colloids over 16 days. Reprinted with permission from A. Peytcheva, H. Cölfen, H. Schnablegger and M. Antonietti, Colloid Polym.Sci. 280,218–227, 2002, Copyright 2002 Springer (ref. [5.28]).

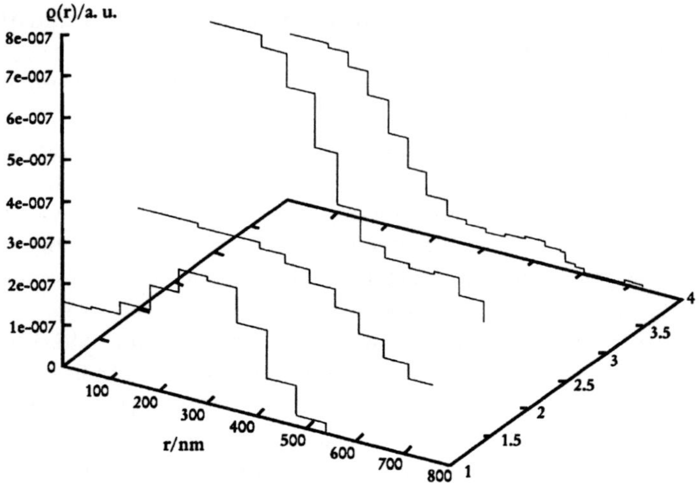

Fig. 5.54. Radial density distributions calculated from the SLS results (Fig. 5.53), using the ORT algorithm and DECON. Reprinted with permission from A. Peytcheva, H. Cölfen, H. Schnablegger and M. Antonietti, Colloid Polym.Sci. 280, 218–227, 2002, Copyright 2002 Springer (ref. [5.28]).

(v) Conclusions:
Numerical algorithms (ORT and DECON) were used to analyze the particle form factors in terms of the corresponding particle density profiles. Although this procedure works fine for SLS from samples containing monodisperse spherical (= isotropic) scattering particles, one should be careful in case of polydisperse or/and nonisotropic particles. For such samples, the numerically obtained radial density profile will become a nonunique solution of the Fourier transform and deconvolution problem. Some approaches on how to treat polydisperse systems or even interacting systems showing structure factor contributions, following this indirect Fourier transform method proposed by Glatter and coworkers, are found in ref. [5.30] and ref. [5.31].

5.3 New Light Scattering Methods

In this last section of Chap. 5, some of the recent technical extensions of the light scattering technique are presented. The first two examples are experimental light scattering setups to overcome the problem of multiple scattering in turbid samples: fiber optic quasi elastic light scattering (FOQELS) and dual color cross correlation. Examples C and D describe the enhancement of the experimental time scale of dynamic light scattering, using a CCD chip as a cheap array detector and as an alternative to more expensive photodiode array detectors (see, for example, the Hamamatsu detector mentioned in example F of Chap. 5.2). Finally, example E will illustrate how an array detector can be used to characterize particles in turbid samples by static light scattering.

Example A (Ref. [2.3]):
Here, I will review the original article by Wiese and Horn which introduced the fiber optic quasielastic light scattering method (FOQELS) to the scientific community.

(i) Samples and sample treatment:
Wiese and Horn studied aqueous polymer latex dispersions with particle size ranging from 41 nm to 326 nm as determined by conventional dynamic light scattering in very dilute solution. The chemical composition was either 100% polystyrene (size 41 nm and 63 nm) or 50% styrene and 50% butylacrylate (115 nm and 199 nm). The largest latex particles of size 326 nm consisted predominantly of styrene with some additional ethylacrylate. All these particles were prepared by emulsion polymerization, using standard surfactants as emulsifiers and $K_2S_2O_8$ as initiator. The latex dispersions were investigated by FOQELS as prepared without further purification. This is a huge advantage of the FOQELS technique, allowing, for example, online monitoring of the average particle size during the synthesis process. Since the dispersions contained a significant amount of salt, corresponding to an estimated ionic strength of 0.025–0.050 mol/L, the Coulomb inter particle interactions were effectively screened at inter particle distances larger than about 5 nm. All aqueous latex dispersions studied by FOQELS had particle concentrations above 1 wt%.

(ii) Light scattering setup:
The scheme of the FOQELS setup has been shown already in Chap. 2.3 (Fig. 2.5), and the experimental principle was described there as well. Here, some more technical details of the method will be given. As determined from the acceptance cone of the optical monomode fiber, the scattering angular range in a typical FOQELS experiment is $180° \pm 3°$, corresponding to a scattering vector $q = (2.640 \pm 0.001) \cdot 10^{-2}\, nm^{-1}$. The authors used a HeNe laser (632.8 nm, 10 mW from Aerotech) as incident light source, a single-mode fiber coupling device (Newport Corporation F-1015) to lock the laser beam into the fiber, and a 4-port, single-mode fiber directional coupler (York SC-633-50-0.5) to separate the back-scattered light from the incident primary beam (where the 4th port was not needed for the FOQELS experiment itself and therefore was submerged into toluene to reduce back reflection of light from the fiber end). The detection unit was a photomultiplier tube (Thorn-EMI PMT 9863) equipped with an amplifier/discriminator (Thorn-EMI C604) and connected to a digital correlator (Brookhaven BI-2030 AT). The optical fiber (= optode) itself was prepared as following: the protective layer was stripped off from the fiber, which then was placed inside a stainless steel tube of diameter 1 mm and sealed afterwards with epoxy resin. The front face of this optode was ground at a slant angle of 10° and subsequently polished with alumina abrasives (grain size 0.3 μm). This grinding

is essential to achieve homodyne detection in the FOQELS DLS experiment, since otherwise interference of the backscattered light with the incident light would cause significant heterodyning. Importantly, the slanting angle of 10° does not affect the scattering angle of 180°.

(iii) Methods used for data analysis:
In concentrated dispersions, the amplitude autocorrelation function includes contributions of the static structure factor, as in ref. [2.3] is expressed by the following equation:

$$g_1(\tau)=\frac{F_s(q,\tau)}{F_s(q,\tau=0)} \tag{5.90}$$

As a consequence, the diffusion coefficient determined from the decay rate of $g_1(\tau)$ is no longer the selfdiffusion coefficient, expected to be independent of q in case of monodisperse spherical particles (see Chap. 1.3.), but becomes the collective diffusion coefficient which depends both on sample concentration and scattering angle.

Here, an important issue is the effect of multiple scattering, usually found in concentrated samples, on the measured autocorrelation function. For FOQELS, it can be shown that decay rates of correlation functions, determined from double and single scattered light, should be equal, and only higher-order (≥ 3) multiply scattered light would cause an additional fast relaxation contribution to the measured correlation function.

Decay rate distributions have been calculated from the measured correlation functions using the CONTIN algorithm. It should be noted once more that, since the particle dispersions studied here were in most cases too concentrated (c >10 g/L), the experimentally determined decay rate distributions do not represent particle size distributions with the exception of the most dilute sample (c = 10 g/L).

(iv) Experimental results:
Figure 5.55 shows the intensity correlation functions measured at different sample concentration, including the free diffusion limit expected at infinite dilution as a reference.

As can be seen in the figure, the decay rates become slower with increasing concentration, and the correlation functions also deviate more and more from a simple single exponential decay. In addition, the data obtained for samples of higher concentration became irreproducible as demonstrated for two selected data sets determined from the samples with the three highest concentrations.

Figure 5.56 shows the decay rate distributions corresponding to the correlation functions presented in Fig. 5.55 as determined by CONTIN analysis.

Alternatively, the data have also been analyzed with the Cumulant method. The mean decay rates determined by these different approaches agree well with the exception of the two samples with the highest concentrations (weight fractions 0.35 and 0.40).

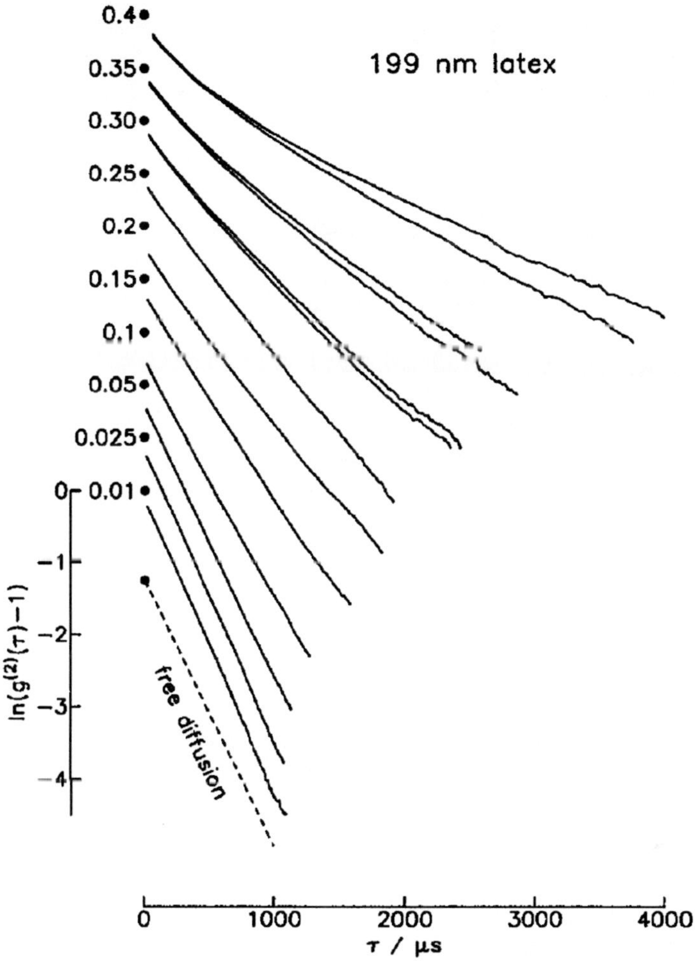

Fig. 5.55. Intensity autocorrelation functions measured by FOQELS for latex particles of diameter 199 nm at different particle weight fractions. The data have been shifted for simpler comparison. Reused with permission from H. Wiese and D. Horn, Journal of Chemical Physics, 94, 6429 (1991), Copyright 1991, American Institute of Physics (ref. [2.3]).

Wiese and Horn further explored the effect of sample concentration on the mean decay rates for latex particles of various particle sizes. Their results, normalized by the decay rates found at infinite dilution, are summarized in Fig. 5.57.

Interestingly, the behavior is nonuniversal, ranging from a strong increase with increasing concentration found for the smaller particles, to a strong decrease found for the larger ones. At 1% particle concentration ($\phi_w = 0.01$), all samples show the behavior expected for highly dilute dispersions, where particle interactions have no influence on particle diffusion. This demonstrates the effective screening of interparticle interactions by the salt present in the samples. The

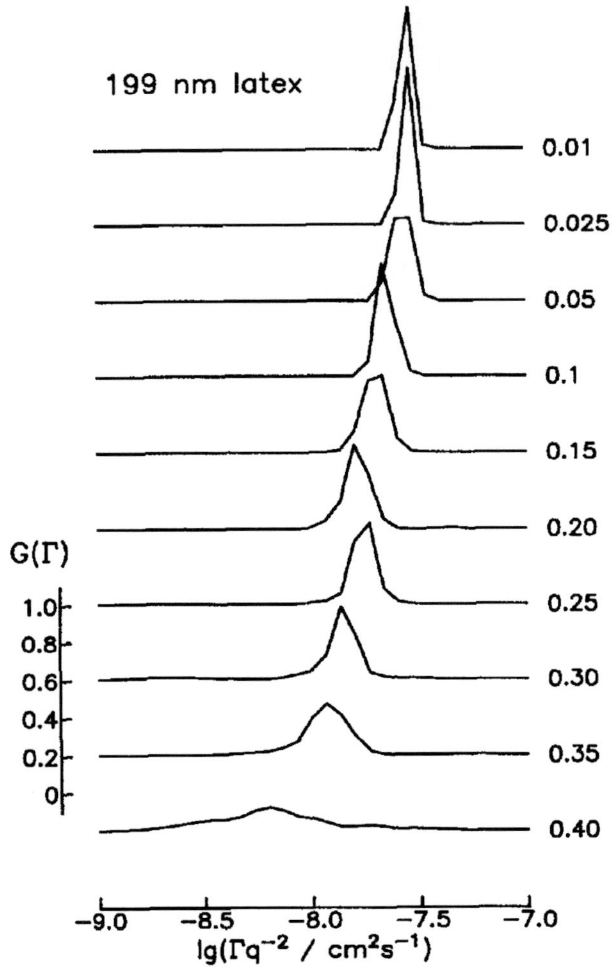

Fig. 5.56. Distribution of FOQELS decay rates computed by CONTIN from the correlation functions displayed in **Fig. 5.55**. Reused with permission from H. Wiese and D. Horn, Journal of Chemical Physics, 94, 6429 (1991), Copyright 1991, American Institute of Physics (ref. [2.3]).

seemingly unexpected effect of particle size on the concentration dependence of the collective diffusion coefficient can be simply explained if we consider the position of the first maximum of the structure factor (= Fourier transform of the particle density distribution) with respect to the scattering vector of the FOQELS experiment. The value q_m, which is a measure for the distance between neighboring latex particles in solution, depends both on particle size and particle concentration, as shown by the simulated data represented in Fig. 5.58.

For q-values larger than q_m, the decay rate of the correlation function usually is interpreted as single particle diffusion. This is plausible since in this

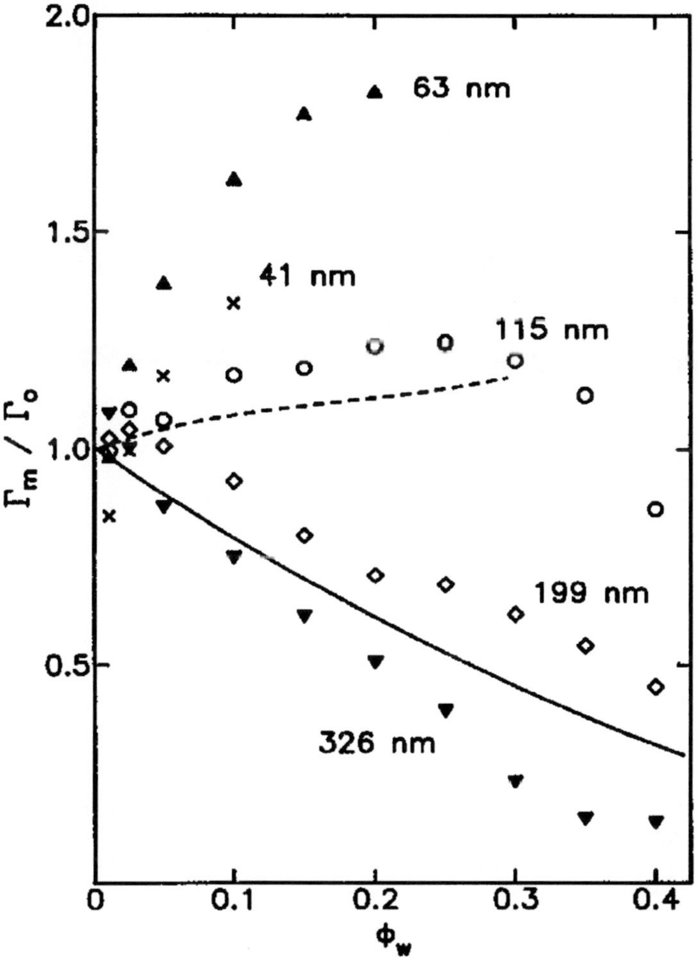

Fig. 5.57. Average FOQELS decay rates versus sample volume fraction ϕ_w for latex particles of various sizes. *Solid* and *dashed lines* correspond to theoretical predictions for the selfdiffusion and the collective diffusion coefficient, respectively, of colloidal hard spheres. Reused with permission from H. Wiese and D. Horn, Journal of Chemical Physics, 94, 6429 (1991), Copyright 1991, American Institute of Physics (ref. [2.3]).

regime the experimental length scale is shorter than the average distance between neighboring particles, and the contribution of interparticle interference to the scattered light can be neglected. In the FOQELS results displayed in Fig. 5.57, this condition is fulfilled for all concentrations of the 326 nm latex and (approximately) the 199 nm latex. Consequently, the relaxation process representing the single particle diffusion becomes slower with increasing concentration, as also theoretically predicted. For latex particles with size 41 nm and 63 nm, $q < q_m$ at all sample concentrations, and therefore the decay rates should represent the collective diffusion coefficient, which is expected to increase with

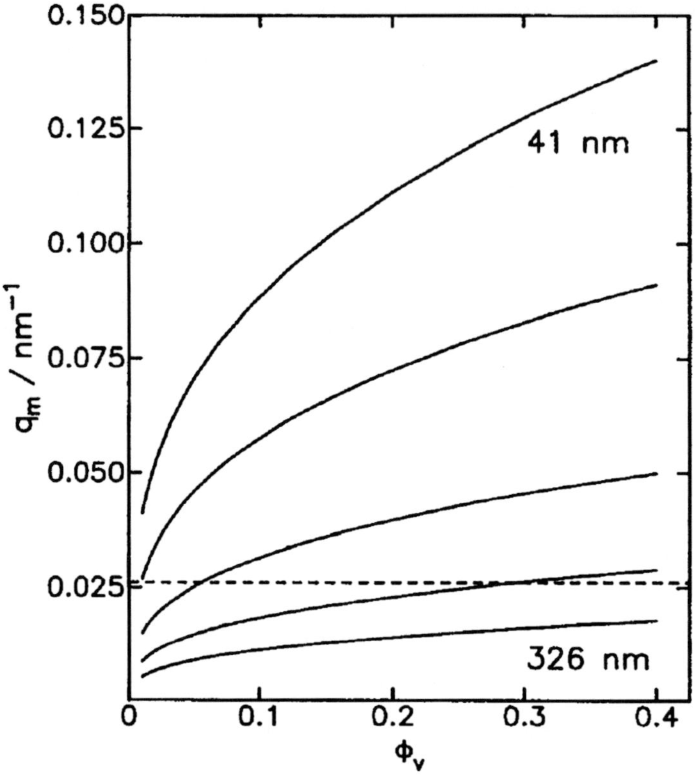

Fig. 5.58. Scattering vector q_m of the first maximum of the static structure factor in dependence of sample concentration for latex particles of various sizes. Reused with permission from H. Wiese and D. Horn, Journal of Chemical Physics, 94, 6429 (1991), Copyright 1991, American Institute of Physics (ref. [2.3]).

increasing concentration. Finally, for the latex with size 115 nm, a mixed behavior is expected depending on sample concentration. Here, Wiese and Horn found a transition from selfdiffusion to collective diffusion for samples with concentrations larger than $\phi_w = 0.1$, as shown in Fig. 5.57.

(v) Conclusions:
The FOQELS technique is a comparatively simple experimental method to successfully suppress multiple scattering, thereby allowing DLS studies of turbid latex samples with concentrations as high as 40 wt%. Due to the very small scattering volume of only about 1e-6 μl (see ref. [2.3]) studied by this technique, compared to a scattering volume of at least 1e-3 μL (= $(0.1\ \text{mm})^3$) for conventional light scattering experiments, purification of the samples from dust usually is not necessary. This makes the method a perfect tool to monitor the sample quality online during the sample synthesis itself. Our example has also shown how to interpret DLS results from concentrated dispersions, where neither the

Stokes–Einstein–equation is valid, nor necessarily single particle diffusion is detected. As a consequence, FOQELS can only be used for single particle analysis or particle sizing if the sample interactions do not yet cause the contribution of a structure factor to the correlation function, which for the salt-screened aqueous latex dispersions studied here seems to be the case for concentrations up to about 1 wt%, as estimated from the concentration dependence of the decay rates shown in Fig. 5.57.

Example B (Ref. [5.32]):
In the previous example, I have reviewed the experimental practice of the FOQELS technique as a dynamic light scattering method to study turbid and highly concentrated colloidal dispersions. Example B will illustrate that dual color crosscorrelation is an alternative by comparing experimental results from both methods for the same sample.

(i) Samples and sample treatment:
Polystyrene latex spheres were prepared by emulsion polymerization of styrene, using commercial poly[styrene-b-(oxyethylene)] block copolymers with molar masses 1,000 g/Mol (PS-block) and 3,000 g/Mol (PEO-block) as emulsifier. The particle diameter, as determined by dynamic light scattering in very dilute solution, was 82 nm, including a stabilization layer of the copolymer chains of thickness approximately 10 nm. This comparatively small particle size assures that for the light scattering measurements always $q \ll q_{max}$, and therefore no selfdiffusion but purely collective diffusion coefficients are measured (see example A). From a comparatively dilute stock latex dispersion, samples with higher solid content up to 25 wt% were prepared by ultrafiltration.

(ii) Light scattering setup:
The FOQELS measurements were made with a home-made device designed as described in example A, using a 5 mW HeNe laser (632.8 nm) as the light source. An ALV-5000 correlator was used to calculate the autocorrelation function of the backscattered intensity fluctuations.

The principle of a dual-color or two-color crosscorrelation (TCC) setup has been described in Chap. 2.4. The two light sources with different wavelengths originated from one single argon ion laser operating in multiline mode, where the laser beams at 488 nm and 514.5 nm were selected as independent light sources. Single-mode optical fibers were used for the optical alignment to avoid instabilities. Scattering angles were varied from 15° up to 140°, showing one major advantage of this technique compared to the FOQELS method: adjustment of the q-scale, whereas the later is limited to a scattering angle of 180°. The light scattering equipment used was a commercial setup from ALV, Langen, and no more technical details are found in ref. [5.32]. Cross correlation functions were determined with an ALV-5000E hardware correlator. As sample cells, standard cylindrical quartz cuvettes with an outer/inner diameter of 10/8 mm immersed in an optical matching vat were used. For samples at concentrations above 8 wt%, a home-made Pyrex glass cuvette with outer/inner diameter of 10/4 mm was used to assure transmission of the incident light through the otherwise too turbid samples.

(iii) Methods used for data analysis:
Correlation functions were analyzed with the commercial ALV software, either by performing the CONTIN inversion or an unconstrained inverse Laplace transformation using the ALV-800 transputer board. Since the ALV software package used by Stieber and Richtering handled directly the intensity autocorrelation function $g_2(q,\tau)$, calculation of the amplitude correlation function $g_1(q,\tau)$ using the Siegert relation had not been necessary for DLS data analysis.

(iv) Experimental results:
Figure 5.59 shows an example for autocorrelation functions and crosscorrela tion functions measured at two different scattering angles, illustrating how suppression of multiple scattering events has a strong influence on the DLS data and especially on the distribution of decay rates representing the correlation function.

Comparing the autocorrelation functions and crosscorrelation functions shown in Fig. 5.59, it is obvious that the contributions from multiple scattering events to the intensity autocorrelation function increase with decreasing scattering angle or increasing length scale. As seen by the CONTIN results displayed in the figure, these multiple scattering events cause a broadening of the decay rate

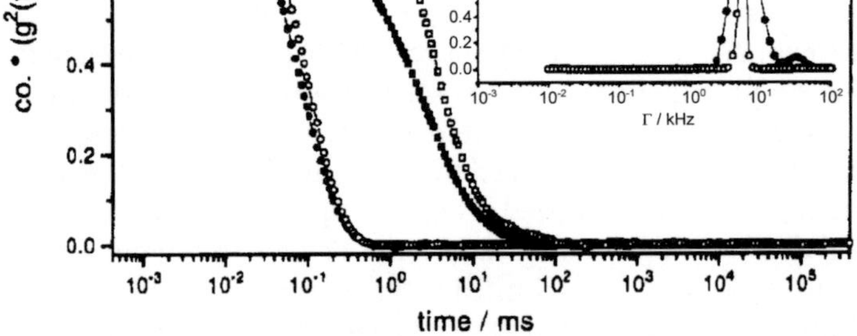

Fig. 5.59. DLS autocorrelation functions (*filled symbols*) and dual color DLS crosscorrelation functions (*open symbols*) for a latex dispersion of c = 0.93% at scattering angles 20° (*squares*) and 125° (*circles*). The inset shows the decay rate spectra obtained by CONTIN analysis. Reprinted with permission from F. Stieber and W. Richtering, Langmuir 11, 4724–4727, 1995, Copyright 1995 American Chemical Society (ref. [5.32]).

distribution, and the appearance of an artificial fast process identified by the additional peak in the decay rate spectrum at higher values of Γ.

At higher particle concentration the crosscorrelation data revealed two relaxation processes. The angular dependence of the two diffusion coefficients extracted from the average decay rates of these two processes is presented in Fig. 5.60.

Results from CONTIN and Laplace inversion shown in Fig. 5.60 are comparable, with the exception that the Laplace inversion can even resolve very small contributions of the slow process (with relative amplitude <0.2) to the crosscorrelation function. The fast diffusion coefficients decrease with increasing scattering vector and agree, at large q, with the result measured independently by FOQELS. On the other hand, the slow diffusion coefficient is nearly independent of the scattering vector, whereas the amplitude of the corresponding relaxation process decreases with increasing q until it becomes undetectable. Since in the experimental q range studied here only the collective diffusion coefficient (see example A) is measured, the q-dependence of the fast diffusion coefficient shown in Fig. 5.60 according to the authors reflects the q-dependence of the static structure factor. With increasing particle concentration, the maximum of the static structure factor is shifted towards larger q corresponding to a decrease in interparticle spacing.

Finally, Stieber and Richtering considered the effect of sample concentration on the two diffusion processes to explore the origin of the second slow diffusive process. Experimental results are shown in Fig. 5.61.

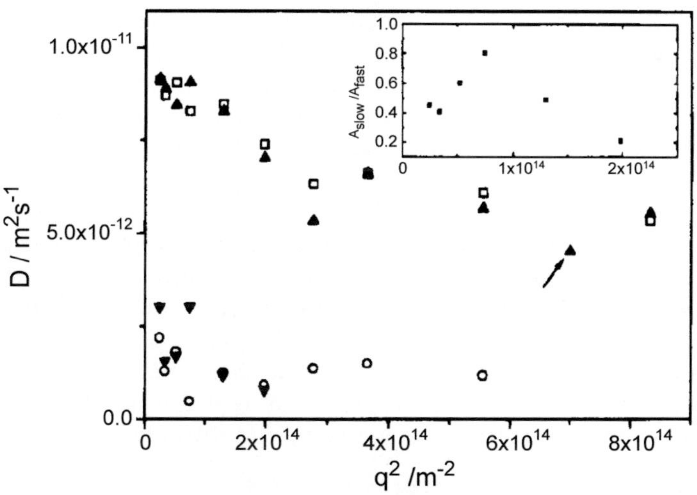

Fig. 5.60. q^2-dependence of the two diffusion coefficients obtained from CONTIN (*filled symbols*) and inverse Laplace Transformation (*open symbols*) of dual color crosscorrelation functions for a latex particle solution with concentration c = 17.6%. The arrow indicates the FOQELS-result, the inset shows the amplitude ratios of the slow and the fast process. Reprinted with permission from F. Stieber and W. Richtering, Langmuir 11, 4724–4727, 1995, Copyright 1995 American Chemical Society (ref. [5.32]).

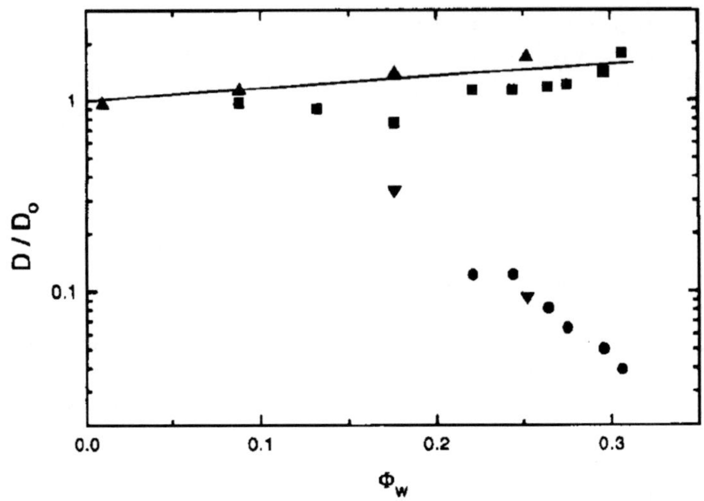

Fig. 5.61. Concentration dependence of the diffusion coefficients (normalized by the Stokes-Einstein diffusion coefficient measured for very dilute samples) as determined by FOQELS (■,●) and dual color crosscorrelation at small scattering angle (▲,▼).Reprinted with permission from F. Stieber and W. Richtering, Langmuir 11, 4724–4727, 1995, Copyright 1995 American Chemical Society (ref. [5.32]).

The fast relaxation process identified as collective particle diffusion increases in velocity with increasing concentration, and values determined from TCC and FOQELS are in good agreement. The slow process appearing at concentrations >17 wt% shows a strong slowing down with increasing concentration and therefore has been identified by the authors as selfdiffusion of the latex particles. The fact that the correlation function represents not only collective particle motion but also single particle motion, according to the authors is caused by an optical polydispersity (= polydispersity of scattering power) of the sample, which itself is based on the size polydispersity of the latex particles. Here, one should recall that the scattering power of a given particle scales with the sixth power of its radius. As a consequence, even a comparatively small size polydispersity of only 10% may lead to a large optical polydispersity. This effect causes an, although minor, contribution of single particle diffusion to the correlation functions measured by DLS of concentrated dispersions, visible here as a second slow relaxation process in the correlation function. It can be shown that the ratio of the amplitudes of the collective diffusion process A_c to that of the selfdiffusion process A_s is given by static structure factor and polydispersity index σ as:

$$\frac{A_c}{A_s} \sim \frac{S(q)}{\sigma^2} \tag{5.91}$$

The polydispersity index is defined as $\sigma = \Delta a / \langle a \rangle$, with Δa the standard deviation of the particle size, and $\langle a \rangle$ the average particle size. Equation 5.91

shows that the amplitude of the slow process should decrease if the value of the structure factor maximum q_m is approached from the small-q regime, which is in agreement with the experimental results shown above (see inset of Fig. 5.60).

(v) Conclusions:
We have seen here that the dual-color method, although technically more complicated than the FOQELS technique due to the difficult optical alignment (since illumination of an identical scattering volume by the two incident laser beams and detection of the scattered light at an identical scattering vector are crucial), has the very important advantage of an adjustable experimental length scale by variation of the scattering angle. This advantage becomes important in case of q-dependent diffusion coefficients, which can be explored in much more detail than has been possible in the previous example A.

Example C (Ref. [2.2]):
The next two examples presented in Chap. 5.3 will illustrate the practical use of CCD camera chips as a cheap detector array in light scattering experiments. Example C shows the application of the CCD device as a simultaneous multiangle detector in the small-q-regime. In this respect, the approach is similar to the multi angle light scattering setup (MALS) described before, although the number of monitored scattering angles typically is much larger in case of the CCD detector (several hundred). In addition, identical q-values are represented by several sectors of the CCD chip, allowing the calculation of an ensemble average and thereby significantly decreasing the total measurement time.

(i) Samples and sample treatment:
The setup has been tested with a very dilute dispersion of commercial latex particles of hydrodynamic diameter 215 nm (Duke Scientific), dispersed in the highly viscous solvent glycerol. The sample was placed in a rectangular cuvette of thickness 1 mm, and kept at a temperature of 51°C during the measurement.

(ii) Light scattering setup:
The light scattering setup based on a CCD camera chip as an array detector, developed by Wong and Wiltzius, is shown in Fig. 5.62.

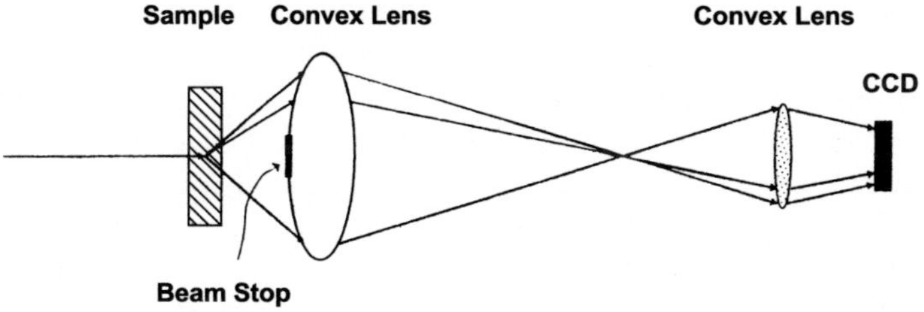

Fig. 5.62. Sketch of a (small-angle) DLS setup using a CCD array detector. Reused with permission from Apollo P.Y. Wong and P. Wiltzius, Review of Scientific Instruments, 64, 2547 (1993), Copyright 1993, American Institute of Physics (ref. [2.2]).

The light source was a 5-mW HeNe laser attenuated to 5e-4 mW in order to use the sensitivity range of the CCD chip in an optimum way. Two convex lenses mapped the light scattered at different scattering angles onto the CCD chip, as shown in Fig. 5.62. Since the lens setup changes the original scattering angles, this mapping was calibrated with the help of a diffraction grating giving rise to a well-defined scattering pattern. The camera was a Pulnix TM845 8-bit CCD with a chip area of $10\times7.2\ \mathrm{mm}^2$, divided into 800×490 pixels^2 with dimension $11.5\times13.5\ \mu\mathrm{m}^2$ each. The optical alignment of this setup was such that all pixels lying on circles around the center of the image corresponded to identical scattering vectors q. The small q range of this experimental setup was comparatively narrow (11,820 cm^{-1}–29,500 cm^{-1}, corresponding to scattering angles of about 4°–10°), potentially limiting its application. Images of the fluctuating scattered intensity, also called "speckle-pattern," were acquired with a Datatranslation DT2851 frame grabber into a PC at video rate and stored in the computer at time intervals of 0.5 s. The computer was programmed to divide the speckle pattern into ten concentric rings each of width 1 pixel such that all pixels on a given ring corresponded to identical scattering vectors with an accuracy of 0.5%. Radii of the subsequent ten concentric rings were 20 pixels to 200 pixels with a step rate of 20 pixels, respectively. The corresponding number of pixels per ring increased from 80 to 800. At each ring, an average speckle intensity $\langle I(q,t)\rangle_{q_0}$ was calculated as well as the deviation of the intensity measured at a given pixel from the average intensity, defined as $\delta I(q,t)=I(q,t)-\langle I(q,t)\rangle_{q_0}$. Note here that the 8 bit-frame grabber used in these studies enables measurement of single pixel intensities as integers between 0 and 255, thereby limiting the accuracy of the method. From the data stored in the computer the time intensity autocorrelation function was calculated according to

$$g_2(q,\tau)=\langle \delta I(q,0)\delta I(q,\tau)\rangle_{q_0} \tag{5.92}$$

For most samples (with the exception of so-called nonergodic systems), this intensity correlation function, which is an ensemble-average over all pixels constituting one predefined ring within the speckle pattern, should be proportional to the usually determined time-averaged autocorrelation function measured at one single q and defined as: $g_2(q,\tau)=\langle I(q,t)I(q,t+\tau)\rangle_t$. The whole process of image acquisition and calculation of the correlation function took about 100 ms in real time, defining the smallest time scale accessible with the experimental method. This limit nowadays could be pushed in principle to shorter correlation times, using a high speed CCD camera, a new generation frame grabber, and a faster computer, whereas the precision of the method could be enhanced by using a 16 bit-framegrabber/CCD-system. For the ten rings, 4,400 correlation functions were calculated in parallel. For comparison, measurements of identical samples with a conventional goniometer-based single-angle light scattering setup using the ALV-5000 hardware correlator, were performed by Wong and Wiltzius.

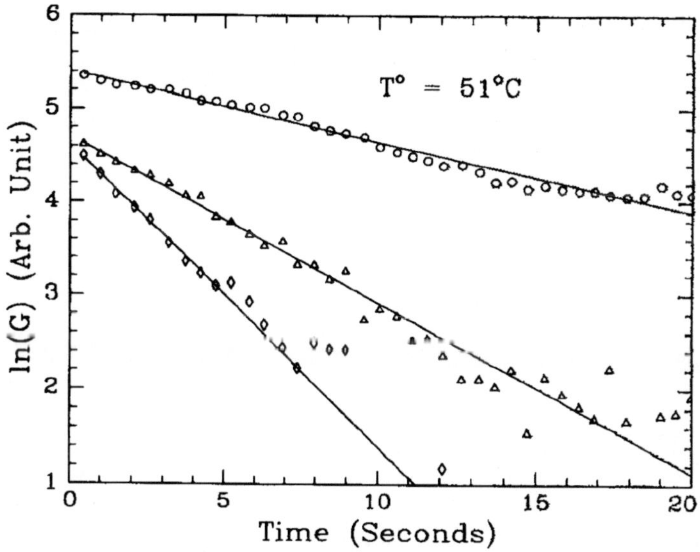

Fig. 5.63. Ensemble-averaged intensity autocorrelation functions determined with the CCD-setup (see **Fig. 5.62**) at three different q-values. Reused with permission from Apollo P.Y. Wong and P. Wiltzius, Review of Scientific Instruments, 64, 2547 (1993), Copyright 1993, American Institute of Physics (ref. [2.2]).

(iii) Methods used for data analysis:
Decay rates were determined by fitting the intensity autocorrelation functions $g_2(q,\tau)$ to single exponential decays.

(iv) Experimental results:
Figure 5.63 shows the ensemble-averaged time intensity correlation functions determined with the CCD-setup at three different scattering vectors (q = 14,832 cm^{-1}, 20,834 cm^{-1}, and 29,748 cm^{-1}), corresponding to three different concentric rings.

Note that the data become increasingly noisy at longer correlation times, but single-exponential data fitting seems to describe the data in a reliable way, indicating the presence of a single relaxation process.

To verify the diffusive character of the single relaxation process and to determine the selfdiffusion coefficient, the authors have plotted the decay rates Γ versus q^2, as shown in Fig. 5.64.

The decay rates plotted in Fig. 5.64 nicely represent the linear dependence on q^2 expected for translational diffusion of single particles in very dilute solution. The slope of the curve yielded a selfdiffusion coefficient of 1.67e-10 cm^{2}s^{-1}, which according to the Stokes–Einstein–equation corresponds to a hydrodynamic particle radius of 107.5 nm, using a measured viscosity of 0.137 Pa s for the solvent glycerol at 51°C. This particle size was in excellent agreement with the latex size given by the commercial provider. Importantly, the result obtained

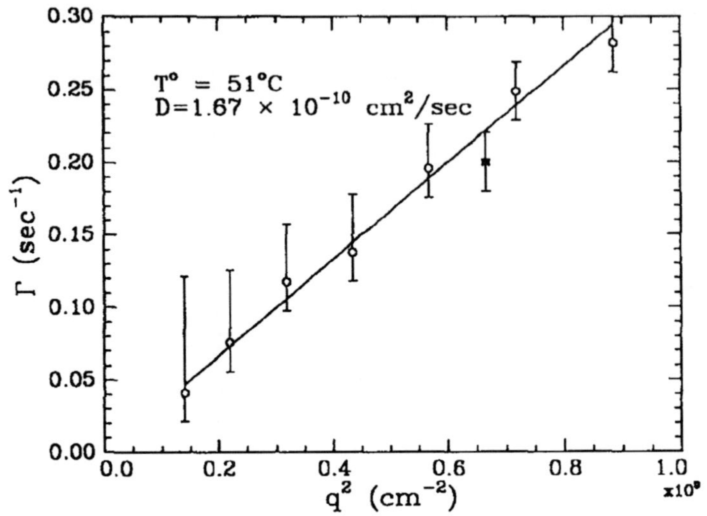

Fig. 5.64. Decay rates of correlation functions measured with the CCD-setup (see **Fig. 5.62**) plotted versus q^2 (*open symbols*), and result determined with a conventional ALV correlator (*filled square*). Reused with permission from Apollo P.Y. Wong and P. Wiltzius, Review of Scientific Instruments, 64, 2547 (1993), Copyright 1993, American Institute of Physics (ref. [2.2]).

from the measurement with the ALV5000 correlator at one scattering angle agreed with the experimental results obtained from the CCD setup within experimental error.

(v) Conclusions:
We have seen a nice example of how a CCD chip and software-based data correlation can replace a photomultiplier and hardware correlator. While the timescale of the CCD setup is much longer than that of conventional DLS setups for technical reasons, the new approach has a major advantage for exploration of slow diffusional processes. This is because of the simultaneous measurement of 4,000 correlation functions, and replacing the time-averaging, which consumes a lot of measurement time corresponding in practice to about 1,000 times the longest correlation time of the autocorrelation function, by ensemble averaging. In this case, the total measurement time and the longest correlation time become identical. This last aspect will be illustrated in more detail by the next experimental example chosen from previous results of the author of this book.

Example D (Ref. [5.33]):
The experimental setup described here is similar to the one just presented in example C, with two important differences: (1) the CCD chip was mounted on the arm of a conventional light scattering goniometer, providing access to the full q range of typical light scattering experiments, and (2) instead of using

a lens setup to project the speckle pattern onto the CCD chip, a pinhole setup was used. All speckles monitored simultaneously corresponded on average to nearly identical q-values, with maximum deviation in scattering angle smaller than ± 0.5°. Correlation functions were calculated from the measured intensities using a mix of ensemble and time-averaging, since our major interest was to shorten the total duration of the light scattering experiment and to enhance the correlation time window compared to conventional setups. The new DLS approach provided experimental access to correlation times >10 s, which is very time-consuming with a conventional DLS setup using only time averaging to determine the correlation function. Importantly, such long-time measurements would afford stability of the optical alignment of the conventional DLS setup for several days.

(i) Samples and sample treatment:
The sample used to check the experimental setup was similar to the one investigated in the previous example C, commercial spherical latex particles of radius 350 nm in a very dilute glycerol solution measured at temperature 10°C. The results were also compared with those obtained by conventional single-angle DLS using an ALV5000 correlator. In addition, the new setup was applied to more interesting dynamic light scattering studies of highly concentrated colloidal dispersions. In this concentration regime, the relaxation functions became too slow to be accurately measured by conventional experimental DLS setups at a reasonable measurement time. A more detailed presentation of these results, however, is beyond the spectrum of this book.

(ii) Light scattering setup:
The light scattering setup, using either a conventional fiber optic detector and hardware correlator ALV5000, or the CCD camera and a software algorithm to calculate the correlation function by combined time- and ensemble averaging with a personal computer, is shown in Fig. 5.65.

Since the CCD chip simultaneously detected a multitude of speckles, the new setup had been called MSCS (multispeckle correlation spectroscopy) in contrast to conventional DLS, which sometimes also is called PCS (photon correlation spectroscopy). The setup consisted of an Ar ion laser (514 nm) attenuated with an adequately turned polarizer to utilize the dynamic range of the CCD chip in its optimum way. The cylindrical light scattering cuvette was immersed in an index-matching bath. The scattered light was passed through a pinhole setup to define the detected q range with an accuracy of ± 0.1°. The CCD camera was a Hitachi KP140 with CCD chip area 11.5×10 mm^2 divided into 580 × 500 pixels^2, and light sensitivity 0.15 Lux. Individual frames of the speckle pattern monitored by the CCD chip were acquired as 8 bit-data with a PC equipped with a Datatranslation DT2255 frame grabber. Images of 512×256 pixels^2 were stored at a frame rate of 0.33 s. To limit the time necessary for data acquisition as well as the amount of data stored, the scattered intensity for a maximum of 50 individual speckles, as defined by areas of 2×2 pixels^2, had been calculated and

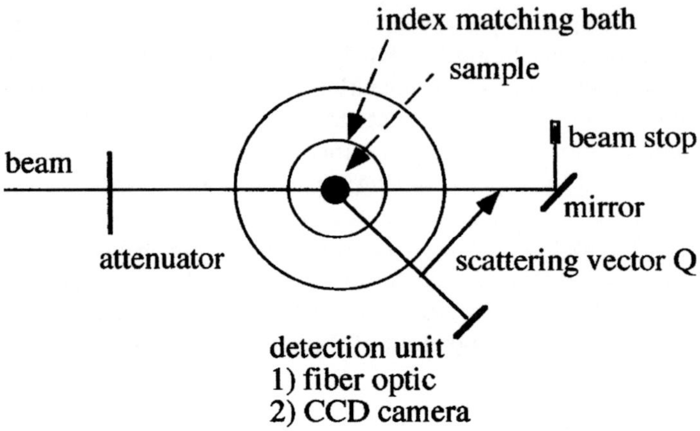

Fig. 5.65. Sketch of a DLS goniometer setup using a CCD-array detector in comparison to the conventional DLS setup. Reused with permission from S. Kirsch, V. Frenz, W. Schärtl, E. Bartsch, and H. Sillescu, Journal of Chemical Physics, 104, 1758 (1996). Copyright 1996, American Institute of Physics (ref. [5.33]).

stored in the computer. These data represented 50 individual traces of the time-dependent scattered intensity $I(t)$, all corresponding to nearly identical scattering vectors. The total amount of data stored was 5 MB as defined by 50 speckles and 100,000 images digitized with 8-bit resolution at successive time intervals. The time- and ensemble-averaged intensity correlation functions were calculated after the data acquisition was complete, using the following expressions:

$$\left\langle I(t)I(t+K\Delta t)\right\rangle_{T,norm} = \\ = \frac{\left(N_{pic}-k\right)\sum\limits_{n=0}^{N_{pic}-k} I(n\Delta t)I(n\Delta t+K\Delta t)-\sum\limits_{n=0}^{N_{pic}-k} I(n\Delta t)\sum\limits_{n=0}^{N_{pic}-k} I(n\Delta t+K\Delta t)}{\sum\limits_{n=0}^{N_{pic}-k} I(n\Delta t)\sum\limits_{n=0}^{N_{pic}-k} I(n\Delta t+K\Delta t)} \quad (5.93)$$

$$\left\langle I(t)I(t+K\Delta t)\right\rangle_{T,E,norm} = \frac{\sum\limits_{a=1}^{N_{sp}}\left\langle I(t)I(t+K\Delta t)\right\rangle_{T,a}}{N_{sp}} \sim \left\langle g_2(q,K\Delta t)\right\rangle_{T,E} - 1 \quad (5.94)$$

First, the normalized intensity correlation function for each individual speckle was calculated by time-averaging according to Eq. 5.93. Δt is the frame sampling time, K an integer value $0 < K < N_{pic}$, defining the correlation time $\tau = K \cdot \Delta t$, with N_{pic} the total number of digitized frames. The time-averaged individual correlation functions next were ensemble averaged over all N_{sp} speckles (Eq. 5.94). The resulting time-ensemble-averaged normalized intensity autocorrelation function had an intercept at $\tau = K\Delta t \rightarrow 0$ smaller than the theoretically expected maximum value of 1.0, depending on the number of coherence areas represented by the observed speckles as already discussed (see Chap. 2). Whereas our standard

DLS equipment uses an optical monomode fiber detector and therefore exhibits an intercept of the normalized intensity correlation function of nearly 1.0 (0.95), the intercept for the CCD setup depends on the pinhole setup and is typically in the order of 0.2. For comparison with the ALV-results, our MSCS-correlation functions therefore had to be shifted by multiplying $\langle I(t)I(t+K\Delta t)\rangle_{T,E,norm}$ with a suitable factor c, that is: $g_2(\tau)_{ALV\,5000}-1=c\cdot\langle I(t)I(t+K\Delta t)\rangle_{T,E,norm}$. Correlation functions of the MSCS setup cover a correlation time range from 0.33 s to more than 3,000 s, whereas the reliable correlation time range of our ALV setup is $10^{-7}\,s<\tau<10\,s$. One should note that correlation functions determined with the ALV5000 hardware correlator show a drastic increase in statistical noise at longer correlation times. For this reason, in experimental practice correlation times $\tau>10\,s$ usually are not accessible with a standard single-angle DLS setup.

(iii) Methods used for data analysis:
The data from ref. [5.33] presented here will not be analyzed in detail. Only the amplitude correlation functions $g_1(q,\tau)$, as calculated from the intensity correlation functions using the Siegert relation, are shown and compared in a semiquantitative way for the two different experimental setups, DLS and MSCS, respectively.

(iv) Experimental results:
Figure 5.66 shows the amplitude correlation functions $f(q,\tau)=g_1(q,\tau)$ measured for the latex/glycerol sample at three different scattering angles, using the MSCS and the DLS-ALV5000 setup.

The MSCS correlation functions here had been determined from an experimental run of duration 50,000 s per scattering angle, resulting in a data set containing 150,000 successive frames of $\Delta t\ = 0.33$ s, each with 20 speckles. For the DLS-ALV5000 measurements, the total experimental time was 70,000 s per scattering angle, which was comparable to the duration of the corresponding MSCS measurement. Importantly, the MSCS data still show a smooth curve at correlation times longer than 50 s. where the DLS-ALV data already become considerably noisy. This benefit of ensemble averaging is also nicely demonstrated by the inset in Fig. 5.66, providing an enlarged detail of the correlation functions in the long-time regime: whereas DLS-ALV data and MSCS results for a single speckle are comparable in noise, averaging of the MSCS data over all 20 speckles leads to a smoothly decaying correlation function.

We also applied the MSCS method successfully to our studies of slow dynamics in highly concentrated colloidal hard sphere dispersions at very long correlation time scales, where conventional DLS not only suffers from noise, but also from insufficient sampling of sample concentration fluctuations. This later effect may lead to erroneously small relaxation times, as shown in more detail in ref. [5.33].

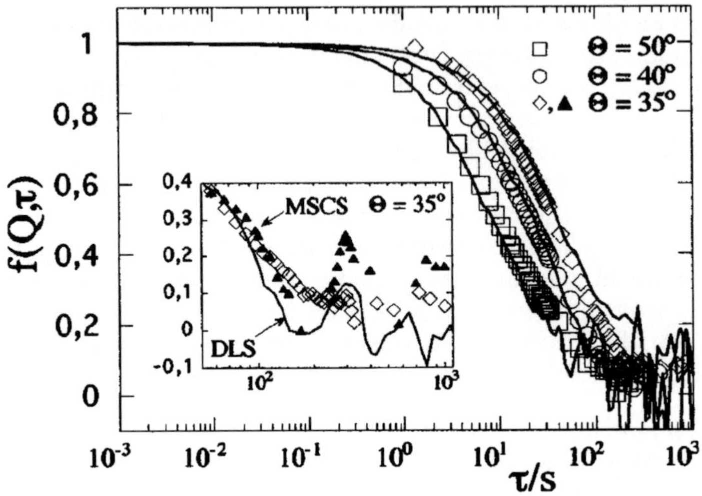

Fig. 5.66. Amplitude correlation functions measured for latex particles dispersed in glycerol at 10°C, using the MSCS-technique (*symbols*) (see **Fig. 5.65**) and the conventional DLS-ALV5000 (*lines*) setup. The inset shows the effect of sample averaging from several simultaneously detected speckles (*open symbols*), compared to the MSCS-correlation function of a single speckle (*filled symbols*). Reused with permission from S. Kirsch, V. Frenz, W. Schärtl, E. Bartsch, and H. Sillescu, Journal of Chemical Physics, 104, 1758 (1996). Copyright 1996, American Institute of Physics (ref. [5.33]).

(v) Conclusions:

Example D underlines the benefits of ensemble averaging using a detector array system, compared to pure time averaging for calculation of the intensity autocorrelation functions in case of a single photomultiplier or photodiode optical detector. The cheapest version of such an array detector could be, as has been demonstrated by two examples, even the CCD chip of a video camera. Recent technical developments, for example high speed cameras, should enable researchers to overcome the major restriction of the CCD-DLS-setups, and thereby provide access to correlation times as short as several μs.

Example E (Ref. [5.34]):

Finally, I will review one example combining the advantages of simultaneous detection at multiple scattering angles (see examples C, D) with an approach to allow particle characterization by static light scattering in case of turbid particle dispersions. Lehner et al. built a new static light scattering instrument, using a very thin, flat light scattering cell in combination with a linear photodiode array detector. This new setup can even be applied to exotic problems not accessible by standard light scattering instrumentation, like the characterization of the average size and size distribution of fat droplets in milk.

(i) Samples and sample treatment:
The performance of the new setup was tested with monodisperse commercial latex particles (Interfacial Dynamics Corporation, Portland, OR) of different particle radii (153 nm, 317 nm, 665 nm, and 1,064 nm). Stock solutions were diluted with filtered doubly-distilled water to adjust particle concentrations of 2.0 e-3 mg/mL in case a standard goniometer light scattering setup was used, and a 1,000-times higher concentration in the case of the new flat cell light scattering equipment.

(ii) Light scattering setup:
As a reference, Lehner et al. used a laboratory-built standard single angle scattering goniometer setup with scattering angular range 6°–150°. The sample cell, a cylindrical cuvette of inner diameter 22.6 mm, or in some cases 8 mm, was embedded in a decalin bath for refractive index matching. This standard setup is useful for transparent samples only, and the single-angle light scattering measurements are rather time consuming. To overcome these limitations, the authors constructed the new flat cell light scattering instrument (FCLSI) shown in Fig. 5.67.

The light source of the FCLSI setup was a 10-mW HeNe laser with random polarization. The light scattering cuvette itself was a very flat sample cell with thickness adjustable between 10 μm and 5 mm, as defined by calibrated distance rings sealed with O rings. This approach enables the suppression of multiple scattering even for very turbid samples. The whole sample cell was embedded in a temperature control jacket connected to a thermostat. Behind the sample cell was a large convex lens with focal length 100 mm. This lens focussed the scattered light onto a photo diode array consisting of 160 individual diodes each of 2 mm width. The

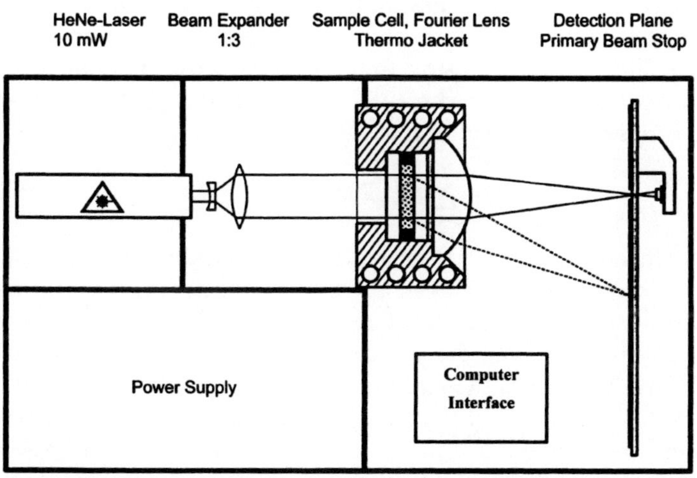

Fig. 5.67. Schematic drawing of a flat cell static light scattering instrument (FCLSI). Reprinted with permission from D. Lehner, G. Kellner, H. Schnablegger and O. Glatter, J. Colloid Interface Sci. 201, 34–47, 1998, Copyright 1998 Academic Press (ref. [5.34]).

angular range of this setup was 1° to 60°. The average scattered intensity detected by all 160 photodiodes was digitized and transferred to a computer with an effective accuracy of 32 bits and a minimum duration of 1 s per readout cycle. To account for the angular dependence of the average scattered intensity, Lehner et al. used a special detection scheme to enhance the dynamic range of their detector array: the acquisition time for photodiodes measuring at the smallest scattering angles (0.4°–10°) was 50 ms, averaged over a period of 1 s via 20 readouts. On the other hand, the acquisition time for photodiodes positioned at the largest scattering angles (42.7°–61.4°) and therefore detecting a much lower scattered intensity, was 1 s per readout. This short measurement time of only 1 s for the complete average scattered intensity $I(q)$ allows time-resolved light scattering characterization of samples not in equilibrium. Finally, inherent different sensitivities of the diodes were taken into account by calibration of the setup with a concentrated small latex particle dispersion used as a scattering standard, leading to respective correction factors for the scattered intensity detected by a given photo diode.

(iii) Methods used for data analysis:
The single scattering average intensity measured in a static light scattering experiment of a noninteracting polydisperse ensemble of spheres with relative refractive index m and radius distribution $D(R)$ according to ref. [5.34] is given as:

$$I(q,m) = \int_{R=0}^{\infty} D(R) W(R) I_1(q,R,m) dR \tag{5.95}$$

$W(R)$ is the so-called weighting function enabling recalculation of the size distribution function into volume (or mass) distributions or the corresponding scattered intensity distributions. $I_1(q,R,m)$ is the particle form factor of a sphere with radius R and refractive index m. In case light absorption is negligible, m is approximately given by the ratio of particle refractive index to solvent refractive index, i. e., $m = n_{D,P} / n_{D,0}$. Using the "indirect transformation technique" developed by Glatter and coworkers [5.35], the size distribution can be transformed into a series of cubic B-splines $\varphi_i(R)$:

$$D(R) = \sum_{i=1}^{N} c_i \varphi_i(R) \tag{5.96}$$

with c_i the unknown coefficients which have to be determined via a least-squares fit. Glatter and coworker developed a software routine which performs this transformation, allowing simultaneous determination of both the refractive index and the size distribution from the static scattering data [5.29]. This routine called "ORT" is available from the authors. Note that a similar mathematical procedure was used by Glatter and coworkers to transform scattered intensities, obtained from monodisperse samples, into radial pair distribution functions, as described in example 5.2 K above. Importantly, this approach (Eqs. 5.95 and 5.96) is only valid in case multiple scattering can be neglected, that is, for light scattering samples with optical transmittance larger than 0.95.

In case of turbid samples or even seemingly transparent samples with optical transmittance lower than 0.95, multiple scattering events have to be taken into account. Lehner et al. expressed the multiply scattered light intensity as a series of Legendre polynomials $P_n(\cos\theta)$:

$$I(z,\theta)=\frac{1}{4\pi}\sum_{n=0}^{\infty}(2n+1)\left[e^{-\mu z(1-B_n)}-e^{-\mu z}\right]P_n(\cos\theta) \tag{5.97}$$,

where z is the path length of the primary laser beam inside the scattering sample, μ is the extinction coefficient of the sample, and B_n contains all information characterizing the single scattering properties of the particles. For the flat sample cells used in the FCLSI instrument, an additional correction factor $f(z,\theta)$ is needed to take into account that here the optical path length varies with scattering angle θ. In conclusion, the measured scattered intensity is given as:

$$I_{\exp}=I(d,\theta)\cdot f(d,\theta)=I(d,\theta)\cdot e^{-\mu d(1/\cos\theta-1)} \tag{5.98}$$

Here, d is the thickness of the flat sample cell. Equations 5.97 and 5.98 were valid as long as the optical sample transmittance was larger than 0.3. Lehner et al. developed the new software package MMS (= multiple Mie scattering), also available from the authors, to determine the volume distribution of the scattering particles from the average scattered intensity $I(d,\theta)$.

(iv) Experimental results:
To illustrate the performance of the new FCLSI as well as the indirect transformation technique, which allows one to obtain particle size distributions of polydisperse samples from SLS measurements, I will briefly review some of the results presented in ref. [5.34]. As not shown here, Lehner et al. first used a Guinier plot ($\log I(q)$ versus q^2) to determine the maximum accessible particle size: they found that the maximum particle diameter which could be measured by the FCLSI is 13.6 μm. Figure 5.68 shows static scattering curves of monodisperse 1,064-nm latex particles obtained from the new instrument and measured by the standard goniometer setup.

The sample thickness was 35 μm for the FCLSI setup and 22.6 mm for the goniometer instrument, allowing for a 1,000-times higher particle concentration in case of the FCLSI although the optical transmission in both cases was above 0.9. The static scattering curves measured by the two different instruments are nearly identical besides a very slight instrumental broadening visible in the maxima and minima of $\log I(\theta)$ detected by the FCLSI, attributed to the finite size of the photo diodes. One important advantage of the FCLSI compared to the goniometer setup is its much smaller scattering angle regime. Calculation of the volume distributions and intensity distributions yielded an average particle size agreeing with that provided by the manufacturer. It should be noted that the volume distribution scales with particle size as R^3, whereas the intensity distribution scales as R^6 (see Chap. 1). If normalized to the same peak height, both distributions should be identical for a given polydisperse sample, making it easy

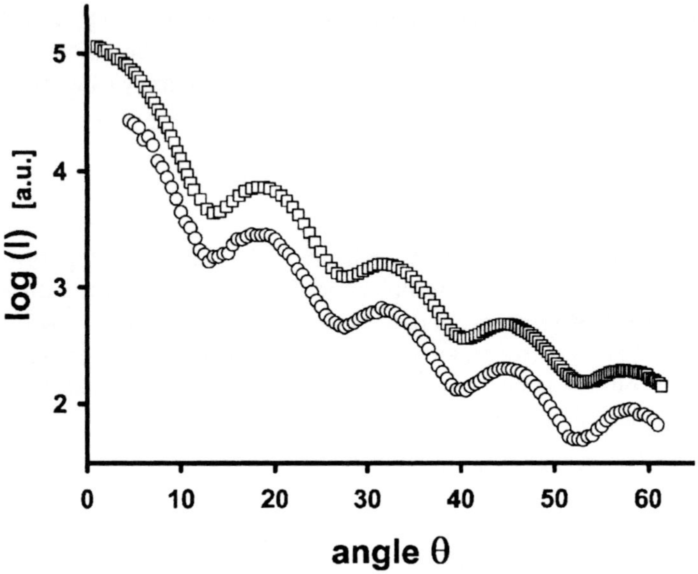

Fig. 5.68. SLS results of monodisperse 1064-nm latex particles measured by FCLSI (upper curve) and a standard goniometer setup (lower curve). Reprinted with permission from D. Lehner, G. Kellner, H. Schnablegger and O. Glatter, J.Colloid Interface Sci. 201, 34–47, 1998, Copyright 1998 Academic Press (ref. [5.34]).

to identify virtual peaks in the distributions determined by the numerical ORT procedure (see ref. [5.34] for more details).

Lehner et al. also investigated polydisperse model samples obtained by mixing equal amounts of latex particles with three different particle radii: 317 nm, 665 nm, and 1,064 nm, corresponding to a size ratio 1:2:3 [see Sect. (ii)]. Importantly, dynamic light scattering would not be able to resolve this trimodal size distribution but only yield a single broad peak for $D(R)$, as has been discussed in Chap. 5.1 in some detail. Figure 5.69 shows the scattering curves detected with the FCLSI for three samples containing only one species of latex samples (c = 1 g/L), and for a trimodal sample containing 1 g/L of each latex particle species.

Figure 5.70 shows the corresponding volume distributions. The trimodal sample composition of the latex mixture is clearly resolved, and the peak positions correspond exactly to those of the respective monodisperse samples. These results demonstrate the potential of the ORT algorithm developed by Glatter et al., in combination with a light scattering instrument covering also the scattering regime of very small scattering angles well below 1°. However, we should consider that the upper scattering angle regime is limited here to $\theta < 60°$. The consequence is that the FCLSI cannot accurately characterize particles of radius smaller than 200 nm, as was demonstrated by Lehner et al. for a second trimodal mixture containing latex particles of sizes 153 nm, 317 nm, and 665 nm. In this case, the distribution peak corresponding to the smallest particle fraction was shifted to 200 nm, clearly

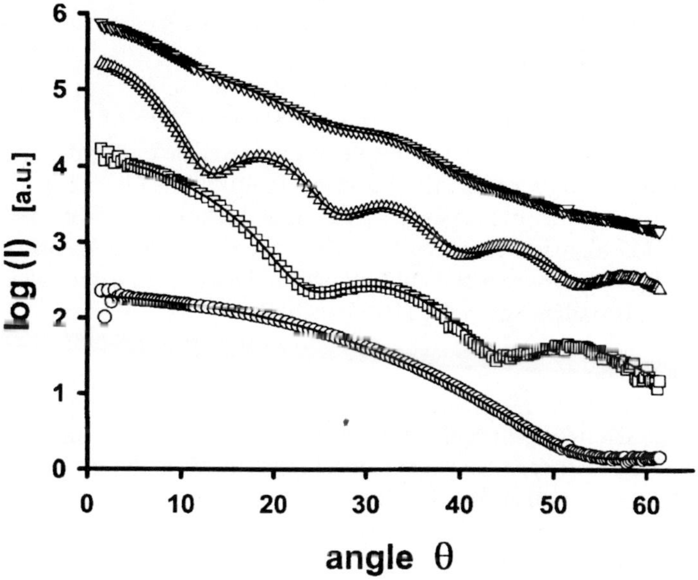

Fig. 5.69. SLS results measured by FCLSI for three different monodisperse latex particles and a 1:1:1 mixture (*upper curve*), see text for sample description. Reprinted with permission from D. Lehner, G. Kellner, H. Schnablegger and O. Glatter, J.Colloid Interface Sci. 201, 34–47, 1998, Copyright 1998 Academic Press (ref. [5.34]).

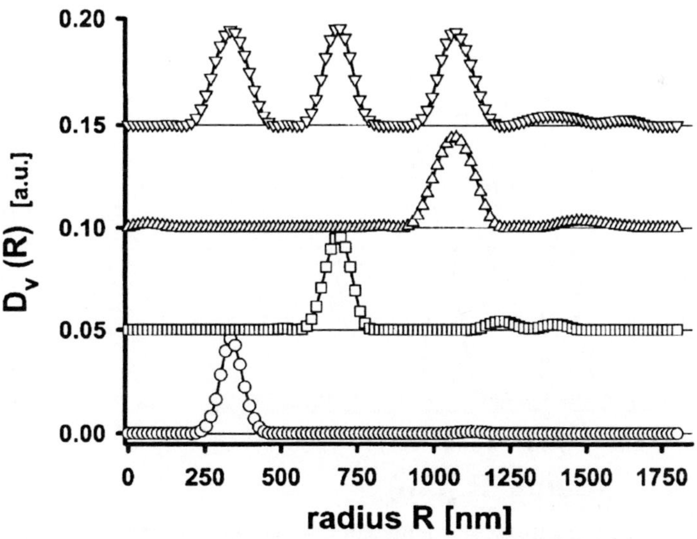

Fig. 5.70. Volume distribution of monodisperse latex samples and a 1:1:1 mixture as calculated from the SLS results (**Fig. 5.69**). Reprinted with permission from D. Lehner, G. Kellner, H. Schnablegger and O. Glatter, J.Colloid Interface Sci. 201, 34–47, 1998, Copyright 1998 Academic Press (ref. [5.34]).

showing the limitations of the FCLSI in combination with the ORT approach to extract the size distribution from $I(\theta)$.

Finally, the authors demonstrated that they could also characterize their trimodal model sample (radii 317 nm, 665 nm, and 1,064 nm) in case of multiple scattering: they investigated a mixed latex sample of total particle concentration 1.5 wt%, that is, five times higher than the previous sample (see Fig. 5.69). As a consequence, the optical transmission even at a sample thickness of 15 μm was only 0.60. The new MMS algorithm yielded a particle size distribution identical to that displayed in Fig. 5.70, whereas the ORT algorithm ignoring multiple scattering caused a strong broadening of the size distribution peaks, as shown in Fig. 5.71.

(v) Conclusions:

Lehner et al. have created a new light scattering setup which, in combination with the highly sophisticated software packages developed in the Glatter group (ORT, MMS, DECON, etc.), enables characterization of particle size distributions by static light scattering even for turbid samples. One major advantage compared to the two DLS techniques described in examples A and B of Sect. 5.3, FOQELS and dual-color DLS, is that the procedure described here can resolve much smaller differences of particle sizes in case of multimodal samples: for the two DLS

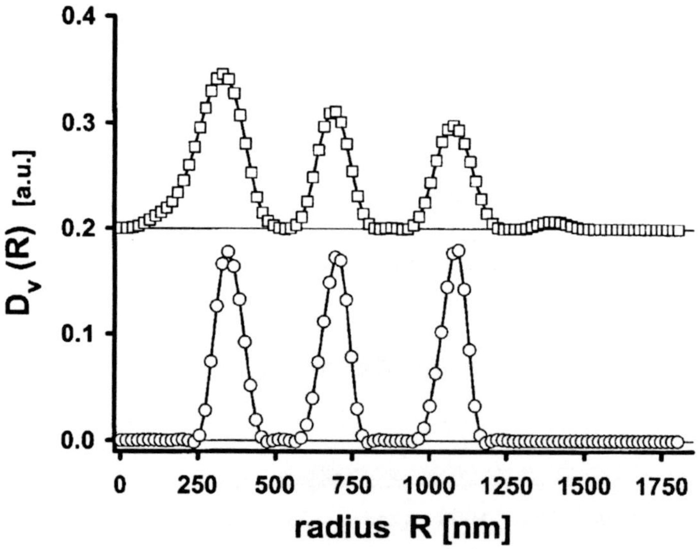

Fig. 5.71. Volume distributions of a tridisperse latex sample (see **Fig. 5.70**) at larger concentration c = 1.5 wt%. The upper curve was obtained from the ORT algorithm ignoring multiple scattering, the lower curve from the new MMS algorithm taking multiple scattering into account. Reprinted with permission from D. Lehner, G. Kellner, H. Schnablegger and O. Glatter, J.Colloid Interface Sci. 201, 34–47, 1998, Copyright 1998 Academic Press (ref. [5.34]).

techniques, the size resolution limit is a ratio of at least 2:1. However, a limitation of the FCLSI is its restriction to rather small scattering angles <60°, and therefore the impossibility of accurately characterizing polydisperse samples containing particles with radii smaller than 200 nm.

References

5.1a Provencher SW (1982) A general purpose constrained regularization program for inverting noisy linear algebraic and integral equations. Comput Phys Commun 27:229

5.1b Provencher SW (1982) A constrained regularization method for inverting data represented by linear algebraic or integral equations. Comput Phys Commun 27:213.

5.2 Galinsky G, Burchard W (1997) Starch fractions as examples for nonrandomly branched macromolecules .4. Angular dependence in dynamic light scattering. Macromolecules 30:6966

5.3 Borsali R, Nguyen H, Pecora R (1998) Small-angle neutron scattering and dynamic light scattering from a polyelectrolyte solution: DNA. Macromolecules 31:1548

5.4 Broersma S (1981) Viscous force and torque constants for a cylinder. J Chem Phys 74:6989

5.5 Tirado M, Garcia de la Torre J (1980) Rotational dynamics of rigid, symmetric top macromolecules. J Chem Phys 73:1986

5.6 Yamakawa H, Fujii M (1973) Translational friction coefficient of wormlike chains. Macromolecules 6:407

5.7 Cong R, Temyanko E, Russo PS, Edwin N, Uppu RM (2006) Dynamics of Poly(styrenesulfonate) Sodium salt in aqueous solution. Macromolecules 39:731

5.8 Lehner D, Lindner H, Glatter O (2000) Determination of the translational and rotational diffusion coefficients of rodlike particles using depolarized dynamic light scattering. Langmuir 16:1689

5.9 van der Zande BMI, Dhont JK G, Bohmer MR, Philipse AP (2000) Colloidal dispersions of gold rods characterized by dynamic light scattering and electrophoresis. Langmuir 16:459

5.10 Ruf H (2002) Treatment of contributions of dust to dynamic light scattering data. Langmuir 18:3804

5.11 Schaertl W, Roos C (1999) Convection and thermodiffusion of colloidal gold tracers by laser light scattering. Phys Rev E 60:2020

5.12 Sehgal A, Seery TAP (1999) Anomalous dynamic light scattering from solutions of light absorbing polymers. Macromolecules 32:7807

5.13 Chu B, Wu C (1987) Light scattering characterization of an alternating copolymer of ethylene and tetrafluoroethylene: 1. Static and dynamic properties. Macromolecules 20:93

5.14 Wu C, Woo KF, Luo X, Ma DZ (1994) A modified light-scattering method for the characterization of the segmented copolymer poly(ethylene terephtalate-co-caprolactone). Macromolecules 27:6055
5.15 Bushuk W, Benoit H (1958) Light-scattering studies of copolymers. I. Effect of heterogeneity of chain composition on the molecular weight. Can J Chem 36:1616
5.16 Schädler V, Nardin C, Wiesner U, Mendes E (2000) Micellization of model macromolecular surfactants as studied by static light scattering. J Phys Chem B 104:5049
5.17 Becker A, Köhler W, Müller B (1995) A scanning Michelson interferometer for the measurement of the concentration and temperature derivative of the refractive index of liquids. Ber Bunsen-Ges Phys Chem 99:600
5.18 Galinsky G, Burchard W (1997) Starch fractions as examples for nonrandomly branched macromolecules. 3. Angular dependence in static light scattering. Macromolecules 30:4445
5.19 Pencer J, Hallett FR (2003) Effects of vesicle size and shape on static and dynamic light scattering measurements. Langmuir 19:7488
5.20 Dobashi T, Narita T, Masuda J, Makino K, Mogi T, Ohshima H, Takenaka M, Chu B (1998) Light scattering of a single microcapsule with a hydrogel membrane. Langmuir 14:745
5.21 Fuetterer T, Nordskog A, Hellweg T, Findenegg GH, Foerster S, Dewhurst CD (2004) Characterization of polybutadiene-poly(ehtylenoxide) aggregates in aqueous solution: A light-scattering and small-angle neutron-scattering study. Phys Rev E 70:1
5.22 Koyama RJ (1973) Light scattering of stiff polymers. J Phys Soc Jpn 34:1029
5.23 Russo PS (1993) Dynamic Light scattering from rigid and nearly rigid rods. Clarendon Press, Oxford
5.24 Wintermantel M, Gerle M, Fischer K, Schmidt M, Wataoka I, Urakawa H, Kajiwara K, Tsukahara Y (1996) Molecular bottlebrushes. Macromolecules 29:978
5.25 Wang XH, Wu C (1999) Light-scattering study of coil-to-globule transition of a poly(N-isopropylacrylamide) chain in deuterated water. Macromolecules 32:4299
5.26 Wu C, Zhou S (1995) Laser light scattering study of the phase transition of poly(N-isopropylacrylamide) in water. 1. Single chain. Macromolecules 28:8381
5.27 Burns JL, Yan YD, Jameson GJ, Biggs S (1997) A light scattering study of the fractal aggregation behavior of a model colloidal system. Langmuir 13:6413
5.28 Peytcheva A, Cölfen H, Schnablegger H, Antonietti M (2002) Calcium phosphate colloids with hierarchical structure controlled by polyaspartates. Colloid Polym Sci 280:218
5.29 Schnablegger H, Glatter O (1993) Simultaneous determination of size distribution and refractive index of colloidal particles from static light-scattering experiments. J Colloid Interface Sci 138:228

5.30 Mittelbach R, Glatter O (1998) Direct structure analysis of small-angle scattering data from polydisperse colloidal particles. J Appl Crystallogr 31:600
5.31 Weyerich B, Brunner-Popela J, Glatter O (1999) Small-angle scattering of interacting particles. II. Generalized indirect Fourier transformation under consideration of the effective structure factor for polydisperse systems. J Appl Crystallogr 32:197
5.32 Stieber F, Richtering W (1995) Fiber-optic-dynamic-light-scattering and two-color-cross-correlation studies of turbid, concentrated, sterically stabilized polystyrene latex. Langmuir 11:4724
5.33 Kirsch S, Frenz V, Schärtl W, Bartsch E, Sillescu H (1996) Multispeckle autocorrelation spectroscopy and its application to the investigation of ultraslow dynamical processes. J Chem Phys 104:1758
5.34 Lehner D, Kellner G, Schnablegger H, Glatter O (1998) Static light scattering on dense colloidal systems: new instrumentation and experimental results. J Colloid Interface Sci 201:34
5.35 Glatter O (1977) A New Method for the Evaluation of Small-Angle Scattering Data. J Appl Crystallogr 10:415

6 Sample Cells, Filters and Solvents

In this chapter, I will briefly review experimental details of light scattering concerning the sample properties and sample preparation itself. First, commercially available sample cells (cylindrical quartz cuvettes) will be presented. Second, commercially available disposable cartridge filters of various pore sizes and suitable for most common solvents used in light scattering will be briefly described. Finally, I will provide a tabular overview of solvent properties needed for quantitative evaluation of light scattering results, containing solvent viscosities at various sample temperatures, solvent refractive indices at room temperature, and refractive index increments dn_D/dc for the most common polymer-solvent-pairs. In case dn_D/dc is not given here for a sample in the reader's interest, it has to be measured from a concentration series of the respective sample which yields the concentration dependent refractive index increment $(n_D(c))$. For this purpose, one of the various commercial refractometers mentioned in Chap. 5 of this book may be used.

6.1 Sample Cells

The company Hellma (see http://www.hellma-worldwide.de/en/default.asp) is one of the most important commercial world-wide providers for suitable light scattering cuvettes. A variety of different cuvettes with inner diameters ranging from 8 mm to 22.6 mm, as shown in Fig. 6.1, are offered:

In practice, larger cuvettes are recommended for a light scattering experiment if possible, since in this case unwanted reflections of the incident and scattered laser beams from the cuvette walls become negligible. The actual operator's choice often depends on the available sample amount. Typically, the sample solution is filtered (see Sect. 6.2) into the light scattering cuvette to a filling height of about 1–2 cm, and the cuvette position in the scattering experiment is adjusted such that the incident laser beam is entering the sample well above the bottom of the sample cell and at least 2 mm below the meniscus of the solution. In this way, the amount of sample is kept comparatively small, while disturbing refractions of the laser beam either from the sample cell bottom or the air-solution interface at the sample meniscus can be avoided. Note that the cuvettes shown in Fig. 6.1 are suitable for most standard light scattering experiments. For some special cases, for example, high temperature light scattering or light scattering combined with

Fig.6.1. Cylindrical light scattering cuvettes, from Hellma, see http://www.hellma-worldwide.de/en/default.asp (courtesy of Hellma)

a fractionation method (GPC, FFF), special sample cells as described in the corresponding examples in Chap. 5 are needed.

6.2 Disposable Syringe Membrane Filters

Disposable cartridge filters provide a simple way to clean most sample solutions from aggregates or dust, which would disturb the light scattering experiment strongly due to the large size of these impurities and their corresponding high scattering intensity. In practice, the solution is filled into a glass syringe and gently squeezed through the filter. Importantly, the first few drops of the sample solution should be discarded during this procedure, since the dry filter membrane may slightly swell or even absorb some of the scattering particles upon contact with the sample solution. Here, an additional aspect is the elution of small particles, which may pollute the scattering sample, from the filters themselves during the early stage of filtration. For static light scattering experiments, where accurate knowledge of the sample concentration is essential, it has to be checked by spectroscopic methods that filtration does not change the amount of scattering solute particles due to the adsorption of sample material at the filter membrane. In case of adsorption, discarding several mL of filtered sample solution before feeding the filtered solution into the light scattering cuvette may solve the problem. In case sample adsorption still cannot be neglected, the actual sample concentration obtained after the filtration process has to be quantified, for example by spectroscopic methods, since it might deviate strongly from the original sample concentration.

Finally, one should keep in mind that during this hand-held filtration procedure high pressures of up to 10 bars may occur (the smaller the syringe the higher the pressure that can be generated: as a general guide, the following pressures can

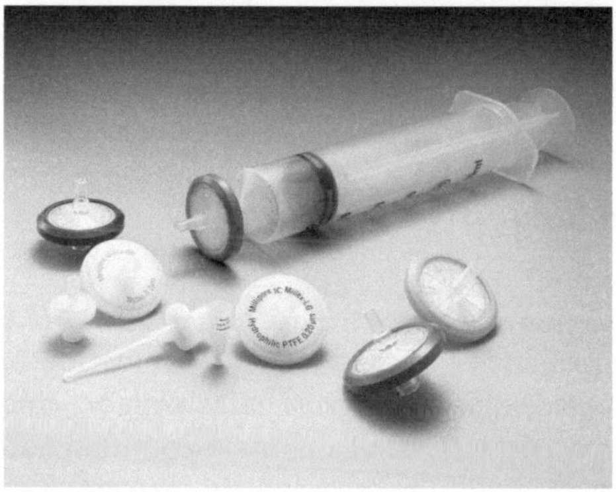

Fig. 6.2. Disposable syringe Millex® filters from Millipore, see http://www.millipore.com (courtesy of Millipore)

be obtained by hand with the syringes indicated: 20 mL, 2 bar; 10 mL, 3.5 bar; 5 mL, 5.3 bar; 3 mL, 7 bar; 1 mL, 10.5 bar). Also, a strong shear of the sample solution while it is squeezed through the membrane has to be considered. As a rule of thumb, the filtration rate should be about one drop of sample solution per second. In case of sensitive samples, as for example micellar aggregates, precautions have to be taken accordingly: waiting for equilibration of the sample after filtration for several hours is one opportunity, and establishment of the equilibrium state can be verified by repeating the light scattering experiment for a given filtered sample several times, in which case the results should not change.

One commercial provider of disposable cartridge filters is the company Millipore (see www.millipore.com). Figure 6.2 shows typical disposable syringe filters type Millex® together with a plastic syringe by hand-held filtration. Note here that for most organic solvents, for example toluene, a glass syringe has to be used.

For filtration of hydrophobic sample solutions, Millipore recommends their disposable filter model "Aervent-50," containing a 50-mm Teflon (PTFE) membrane with pore size 0.2 μm. The maximum inlet pressure for these filters is 4.1 bar. At higher pressure the membrane and/or the filter housing could be destroyed. The filters are suited for most common organic solvents with the exception of pure hydrocarbons (e. g., benzene, toluene, xylene, hexane, etc.), where limited use is specified. For filtration of aqueous solutions, Millipore provides their color-coded Millex® filters with three different types of membranes: millipore express (PES), mixed esters of cellulose (MF), and Durapore (PVDF). Pore sizes available are 0.22 μm, 0.45 μm, and 0.8 μm. The smallest pore size is needed to obtain a sterile aqueous sample solution, that is removal of both dust and "biological impurities" (bacteria, algae, etc.), whereas the larger pore sizes are for prefiltration only.

An alternative commercial provider of disposable syringe membrane filters is the company Whatman (for more information, see http://www.whatman.com/). Their Syrfil® MF disposable syringe filters with a diameter of 25 mm can be used both for aqueous and organic solutions and are equipped with Membra-Fil cellulosis membranes in a Cryolite® housing, including a unique flow deflector that enhances flow and reduces back pressure. Pore sizes available again are 0.22 μm, 0.45 μm, and 0.8 μm.

6.3 Characteristics of Common Solvents

In this last section of the book, as mentioned above, the reader will find two compiled lists of characteristic solvent properties needed for data evaluation of light scattering results, such as solvent viscosity η, index of refraction n_D and refractive index increment dn_D/dc, for a selected list of the most common solvents and polymer-solvent-pairs. Let us start with the pure solvent properties η and n_D listed in Table 6.1.

With the parameters P_1, P_2 provided in the table, the solvent viscosities (in centipoise [cp]) can be calculated for different sample temperatures using the following equation:

$$\eta(T/K)/cp = 10^{(P1+P2\cdot 1000/T)} \tag{6.1}$$

$\eta(T)$ is needed for the quantitative evaluation of dynamic light scattering experiments, since the viscosity of the solvent may vary strongly with sample temperature. On the other hand, the temperature dependence of the refractive index dn_D/dT can usually be neglected in the typical experimental temperature range of 5°C to 80°C. Note that the lower sample temperature in a standard light scattering experiment is limited due to condensation of water vapour from the environment on the glass surfaces of sample cell or index matching bath and optical components in case no special climate-controlled laboratory is used. On the other hand, the upper temperature limit is given by the boiling point of the solvent or the index matching bath liquid.

For the evaluation of static light scattering results, in many cases (for example, determination of the molar mass of the scattering particles using the Zimm equation, see Chap. 1.2 and Eqs. 1.12, 1.21) absolute scattered intensities are needed. These values in practice are determined using a pure solvent like toluene as scattering standard. The absolute scattered intensity of pure toluene is 1.39 e-5 cm^{-1} at an irradiation wavelength of 632.8 nm and a sample temperature of 20°C [6.1]. In addition to the absolute scattered intensities, the Zimm-analysis of static light scattering data also needs the optical contrast factor K as input parameter. This quantity depends on the wavelength of the incident irradiation, the refractive index of the solvent and, importantly, the refractive index increment dn_D/dc (see Eq. 1.9). Chapter 6 concludes with a list of dn_D/dc-values for selected polymer-solvent-pairs taken from ref. [6.1] (see Table 6.2).

The reader should note that both wavelength of the incident light and sample temperature have an effect on dn_D/dc. For a comprehensive list of dn_D/dc-values the interested reader is referred to ref. [6.1].

Table 6.1. Solvent properties (viscosity η and refractive index n_D), all data from Sigma-Aldrich (www.sigma.com), parameters P1, P2 to calculate the temperature-dependent viscosities with courtesy Dr. Karl Fischer, Mainz University

Solvent	Formula	$\eta(20°C)/cp$	P1	P2	$n_D(20°C)$
Acetone	C3H6O	0.32	−1.70219	0.35538	1.359
Acetonitrile	C2H3N	0.37	−1.72057	0.37500	1.344
Benzene	C6H6	0.65	−2.02031	0.53697	1.501
1-Butanol	C4H10O	2.95	−3.01053	1.01872	1.399
n-Butyl acetate	C6H12O2	0.73	−1.94812	0.53145	1.394
t-Butylmethylether	C5H12O	0.27	−1.80088	0.39438	1.369
Chloroform	C1H1Cl3	0.58	−1.55476	0.38438	1.446
Cyclohexane	C6H12	0.98	−2.22196	0.64856	1.426
Cyclopentane	C5H10	0.44	−1.78275	0.41707	1.400
1,2-Dichlorobenzene	C6H4Cl2	1.32 (25°C)	−1.69081	0.53810	1.551
1,2-Dichloroethane	C2H4Cl2	0.79	−1.82127	0.51128	1.445
Diethylether	C4H10O	0.24	−1.81166	0.34615	1.353
N,N-Dimethylacetamide	C4H9NO	2.14	−2.07404	0.60932	1.438
N,N-Dimethylformamide	C3H7NO	0.92	−1.75118	0.49525	1.430
1,4-Dioxane	C4H8O	1.44 (15°C)	−2.16276	0.66765	1.422
Ethanol	C2H6O	1.10 (25°C)	−2.35086	0.71270	1.363
Ethyl acetate	C4H8O2	0.46	−1.82922	0.43481	1.372
n-Hexane	C6H14	0.31	−1.84333	0.39064	1.375
Methanol	C1H4O	0.55	−2.00929	0.52115	1.329
1-Propanol	C3H8O	2.26	−2.86405	0.94125	1.384
2-Propanol	C3H8O	2.86 (15°C)	−3.52978	1.14587	1.377
Pyridine	C5H6N	0.95	−1.94905	0.56612	1.510
Tetrahydrofurane	C4H8O	0.55	−1.66558	0.39721	1.407
Toluene	C7H8	0.59	−1.74089	0.44318	1.496
Water	H2O	1.00	−3.04585	0.89372	1.333

Table 6.2. Refractive index increments for selected polymer-solvent-pairs (see also ref. [6.1])

Polymer-solvent	dn_D/dc (25°C, 436 nm) (mL/g)	dn_D/dc (25°C, 546 nm) (mL/g)
Bovine serum albumin – 0.2 M acetate buffer	0.193	
Bovine serum albumin – phosphate buffer (pH 7)	0.189	
Bovine serum albumin – water	0.195	
Dextran – water	0.154	0.151
Ludox – diluted with water	0.062	0.061
Polybutadiene – cyclohexane	0.126	
Polybutadiene – THF		0.132
Polydimethylsiloxane – toluene	–0.103	–0.093
Polyethylene glycol (M=6000) – methanol		0.149
Polyethylene glycol (M=6000) – water	0.145	0.135
Polyisoprene (70% cis 1,4) – cyclohexane		0.103 (23°C)
Polyisoprene (cis 1,4) – cyclohexane		0.128 (20°C)
Polymethylmethacrylate – acetone	0.131	0.129
Polymethylmethacrylate – benzene	0.004	0.011
Polymethylmethacrylate – dioxan	0.072	0.071
Polymethylmethacrylate – DMF	0.060 (20°C)	0.061 (20°C)
Polymethylmethacrylate – THF	0.089	0.087
Polymethylmethacrylate – toluene	0.009	0.016
Polystyrene – chloroform		0.169
Polystyrene – cyclohexane	0.181	0.170
Polystyrene – dioxan	0.185	0.173
Polystyrene – MEK	0.232	0.223
Polystyrene – toluene	0.115	0.111
Polystyrene (latex) – water	0.257	
Polyvinylacetate – water	0.168 (30°C)	0.164 (30°C)
Poly-4-vinyl pyridine – DMF	0.160	
Poly-4-vinyl pyridine – methanol	0.267	
Tobacco mosaic virus – water	0.176	0.170

References

6.1 Huglin MB (ed) (1972) Light scattering from polymer solutions. Academic Press, London

7 Further Reading

Although in the preceding six chapters I tried my best to provide the reader with a thorough understanding both of theory and practice of modern light scattering, I will not claim to have covered all important aspects of the method within the context of this comparatively short laboratory handbook. For readers interested in more details or special applications of the light scattering technique not treated here, I therefore conclude my book with a list containing some of the (up to now) most important books about light scattering, including their respective tables of contents.

(A) Scattering of Light & Other Electromagnetic Radiation
by Milton Kerker
Publisher: Academic Press (July 1997)
Brief description:

This is the newest available edition of the standard textbook on light scattering theory.

Table of Contents:

(B) Light Scattering by Small Particles (Structure of Matter Series.)
by H. C. van de Hulst
Publisher: Dover Publications (December 1981)
Brief description:

This text offers a comprehensive treatment of the light-scattering properties of small, independent particles, covering both basic scattering theory and particular computations with different kinds of particles. It includes a full range of useful approximation methods for researchers in chemistry, meteorology, and astronomy.

Table of Contents:

1. Introduction
2. Conservation of energy and momentum
3. Wave propagation in vacuum
4. Wave propagation in a medium containing scatterers
5. Polarized light and symmetry relations
6. Particles small compared to the wavelength
7. Rayleigh–Gans scattering
8. Particles very large compared to the wavelength
9. Rigorous scattering theory for spheres of arbitrary size (Mie theory)
10. Non-absorbing spheres
11. Spheres with refractive index near 1
12. Very large spheres
13. Optics of a raindrop
14. Absorbing spheres
15. Circular cylinders
16. Particles of other forms
17. Edge phenomena and surface waves
18. Scattering and extinction experiments as a tool
19. Applications to chemistry
20. Applications to meteorology
21. Applications to astronomy

(C) Dynamic Light Scattering: With Applications to Chemistry, Biology, and Physics
by Bruce J. Berne, Robert Pecora
Publisher: Dover Publications; Unabridged edition (August 2000)
Brief description:

This is a new paperback edition of the classic book on dynamic light scattering by Berne and Pecora which first appeared in 1975. It provides a comprehensive introduction to the principles underlying laser light scattering, focusing on the time dependence of fluctuations in fluid systems. The book also serves as an introduction to the theory of time correlation functions, with chapters on projection operator techniques in statistical mechanics. Wide-ranging discussions of numerous applications make this volume of interest to research chemists, physicists, biologists, medical and fluid mechanics researchers, engineers, and grad students.

Table of Contents:

(D) Light Scattering from Polymer Solutions (Physical Chemistry, a Series of Monographs)
M.B. Huglin (Editor)
Publisher: Academic Press Inc., USA (August 1972)
Brief description:

This book is the classic text on static light scattering, written mostly before the widespread introduction of lasers, personal computers, and the dynamic light scattering technique. It is full of valuable data and insight but is long out of print.

Table of Contents:

(E) Dynamic Light Scattering: The Method and Some Applications
Wyn Brown (Editor)
Publisher: Oxford, Clarendon Press (1993)
Brief description:

This book is a comprehensive treatment of the philosophy and practice of dynamic light scattering. It illustrates the widely varying fields in which the technique finds application: chapters address multicomponent mixtures, polyelectrolytes, dense polymer systems, gels, rigid rods, micellar systems, and the application of dynamic light scattering to biological systems.

Table of Contents:

1. Dynamic scattering from multicomponent polymer mixtures in solution and in bulk
2. Single-photon correlation techniques
3. Noise on photon correlation functions and its effect on data reduction algorithms
4. Data analysis in dynamic light scattering
5. Dynamic light scattering and linear viscoelasticity of polymers in solution and in the bulk
6. Dynamic properties of polymer solutions
7. Application of dynamic light scattering to polyelectrolytes in solution
8. Simultaneous static and dynamic light scattering: application to polymer structure analysis
9. Dynamic light scattering from dense polymer systems
10. Dynamic light scattering from polymers in solution and in bulk
11. Dynamic light scattering from polymer gels
12. Dynamic light scattering from rigid and nearly rigid rods
13. Light scattering in micellar systems
14. Critical dynamics of binary liquid mixtures and simple fluids studied using dynamic light scattering
15. Application of dynamic light scattering to biological systems
16. Diffusing-wave spectroscopy

(F) Light Scattering: Principles and Development (Monographs on the Physics and Chemistry of Materials)
Wyn Brown (Editor)
Publisher: Oxford University Press, USA (August 1996)
Brief description:

This book is directed to the recent development in the light scattering technique and to describing a wide spectrum of its applications. Both the theoretical development and utilization are traced by authors who are experts in their fields.

Table of Contents:

(G) Particle Characterization: Light Scattering Methods (Particle Technology Series)
by Renliang Xu
Publisher: Springer; 1st edition (January 2002)
Brief description:

This very recent book describes in detail the modern practice of particle sizing and particle counting. In addition, a summary of all major particle sizing and other characterization methods, basic statistics and sample preparation techniques used in particle characterization, as well as almost 500 latest references are provided.

Table of Contents:

Index

MIX
Papier aus verantwortungsvollen Quellen
Paper from responsible sources
FSC® C105338

If you have any concerns about our products,
you can contact us on
ProductSafety@springernature.com

In case Publisher is established outside the EU,
the EU authorized representative is:
Springer Nature Customer Service Center GmbH
Europaplatz 3, 69115 Heidelberg, Germany

Printed by Libri Plureos GmbH
in Hamburg, Germany